# Twitterbots

# Twitterbots

## Making Machines that Make Meaning

Tony Veale and Mike Cook

The MIT Press
Cambridge, Massachusetts
London, England

This book was set in ITC Stone Serif Std by Toppan Best-set Premedia Limited. Printed and bound in the United States of America.

Library of Congress Cataloging-in-Publication Data

Names: Veale, Tony, 1967- author. | Cook, Mike, author.
Title: Twitterbots : making machines that make meaning / Tony Veale and Mike Cook.
Description: Cambridge, MA : The MIT Press, [2018] | Includes bibliographical references and index.
Identifiers: LCCN 2017050462 | ISBN 9780262037907 (hardcover : alk. paper)
Subjects: LCSH: Twitter. | Twitterbots. | Natural language generation (Computer science)
Classification: LCC HM743.T95 V43 2018 | DDC 006.3/5--dc23 LC record available at https://lccn.loc.gov/2017050462

10  9  8  7  6  5  4  3  2  1

# Contents

# 1 Less Is More

## Educated Insolence

In an age when people still sent telegrams and paid for their messages by the word, the telephone companies—which were not to be outdone by an older technology—would proudly proclaim, "Every telephone is a telegraph office."[1] Much like today's mobile phones, this combination of technologies made the world a smaller and more connected place, allowing people to do many of the same things that web-savvy users go online to do today, such as transferring money, booking tickets for passage by rail or sea, and ordering flowers, candy, books, and cigars for delivery to recipients in cities across the globe. In the late nineteenth and early twentieth centuries, when the world was connected not by the Internet or the web but by transatlantic cable, and the "last mile" was just a boy on a bicycle, how-to books such as Nelson Ross's "How to Write Telegrams Properly," a 1928 pamphlet, would joke that "brevity is the soul of telegraphy." Though it is often said that it costs nothing to be polite, the telegraph was a communication medium whose users had to pay to say "please" and pay twice as much again to say "thank you." If the telegraph was the Internet of its day, the telegram was its tweet.

We tend to use words sparingly when we have to buy them retail. Yet while constraints often bring out the best in us, telegraphy was not widely considered a medium in which writers did their best work. Even James Joyce, one of the most creative writers of the twentieth century, could only muster the three-word missive "son born Jim" to his brother Stanislaus on the birth of his son Giorgio.[2] Ernest Hemingway's editor, Maxwell Perkins, was even briefer, sending a one-word telegram, "Girl," on the birth of his daughter, while Sigmund Freud joked—in a way that will surprise no one—that while news of a boy surely deserved a telegram, news of a girl warranted only a letter.[3]

Telegrams were much faster than letters, of course, but they discouraged verbosity and encouraged instead a system of conventions, shorthands, and codes. However, even conventions are open to playful exploitation. When physicist Edward Teller, not a man celebrated for his linguistic creativity, telegraphed colleagues to notify them of the first successful detonation of a hydrogen bomb, his telegram was not unlike Joyce's: "It's a Boy."[4] Yet we should perhaps expect more from writers who are paid to be funny. When dispatched to Venice on an assignment by his editor at the *New Yorker*, the humorist Robert Benchley sent off this six-word telegram, "Streets full of water. Please advise," which established the high-water mark of wit in the medium.[5] But such stories are very much the exception rather than the norm, since telegrams were rarely intended for mass consumption. We know only of the best because their authors chose to share them after the fact, in anecdotes that improved with age. In some tellings, Benchley omitted the extravagance "please," and in others he replaced "full of water" with "flooded." Yet the best examples of the medium cost nothing at all to send because they were never actually sent, except in that world of third-hand anecdotes and after-dinner speeches where fable becomes fact. Sadly, the telegram that is often cited as the wittiest ever written belongs to the realm of the never sent. It would count as just another example of staircase wit from the age of telegraphy if the year it was supposed to have been sent, 1843, was not a full year before Samuel Morse sent the first official telegram—the grandiose "What God Hath Wrought"—in 1844.[6]

When General Charles Napier conquered the Indian province of Sindh (now part of Pakistan) in 1843 on behalf of his employers in the East India Company, he is said to have cheekily sent them the one-word telegram, "*Peccavi*."[7] Napier was a veteran of the Peninsular War, whose mandate in Sindh was to suppress the rebellious elements who were making commerce difficult for his employers, and he brought with him the public school philosophy that one is never more predisposed to gratitude than after receiving a sound thrashing. But Napier exceeded his mandate by brutally bringing the whole province to heel, and though he was richly rewarded for his efforts, his military zeal was the target of much criticism in the newspapers and in parliament. So *Peccavi*, Latin for "I have sinned," was both a confession of his guilt and a celebration of his military victory—in other words, "I may have sinned *but* I have Sindh." It didn't hurt that Napier could show off his classical education in the process and flatter his employers too by acknowledging that they were also educated enough to understand Latin.

Napier's witticism rings true mainly because we want it to be true. It caps a witty anecdote that hides the horrors of imperial repression behind a clever pun. Like the many instances of verbal ineptitude that folk history attaches to Vice President Dan Quayle—such as the tale of how, on a tour of Latin America, Quayle expressed regret at not having taken Latin classes in high school—we prefer the humorous legend to the boring truth. But just as Quayle's tale of Latin witlessness was invented by a late-night comic and later misremembered and misquoted as historical fact by a willing electorate, Napier's tale of Latin wit was invented by a schoolgirl, Catherine Winkworth, who joked to her teacher that *"Peccavi"* would have been the wittiest way for the overzealous general to signal his triumph to his disapproving bosses.[8] Winkworth sent her joke to the editors of a new humor magazine, *Punch*, and *Peccavi* soon became part of the official Napier biography.

If Napier's telegram that never was seems like a lost opportunity to impress future historians, future generals were more than ready to make up for Napier's oversight. In 1856, when the British annexed the Indian province of Oudh (whose name rhymes with *loud*), the governor-general of India, Lord Dalhousie, sent a one-word telegram, *"Vovi,"* to the Foreign Office in London. Taking his cue from the schoolgirl of the decade before, Dalhousie's *Vovi* is Latin for "I have vowed," and can thus be read as a pun for "I have Oudh [as I vowed]." But the governor's annexation of Oudh was to stir rebellious feelings among the ill-treated population, forcing General Colin Campbell (later Baron Clyde) to take and later retake the city of Lucknow following the Sepoy mutiny of 1857. After capturing Lucknow for the second time in 1858, Campbell is said to have sent yet another Latin joke by telegraph. But inflation was clearly taking its toll, for Campbell now needed three words—*"Nunc fortunatus sum,"* meaning "I am in luck now"—to signal his victory with a pun. In military circles, the Latin pun was fast becoming a telegraphic meme with which generals could simultaneously paint a veneer of polite society over the brutality of imperialism *and* cement their reputations in the history books.[9]

We can see why puns such as *"Peccavi"* and its later variants might have appealed to a clever child such as Catherine Winkworth. Like most other instances of creativity, linguistic or otherwise, Winkworth's pun made the strange seem more familiar and the familiar seem just a little stranger and more exotic. New ways of referring to the distant corners of the British Empire could be fashioned from the stuff of everyday schoolwork, while the banal substance of this work—*boring Latin*—could be put to new and clever uses. But what could drive such men of state as Dalhousie, and such men of war as Napier and Campbell, to quite literally speak (if only in our collective imaginations) like a schoolgirl, albeit one with a classical education and a dry wit? Aristotle said it best when he defined humor as a form of educated insolence, for linguistic creativity is an essentially precocious aspect of the way we use language.[10] Its precocity is anchored in a number of seemingly contradictory desires: the desire to fit in, balanced with the desire to show off; the desire to respect tradition while demonstrating a mastery over convention; the desire to belong while striving to stand out; and the desire to follow (or be seen to follow) in the footsteps of past masters while blazing a trail of one's own. Having expended blood, treasure, and goodwill to secure a brutal victory, it may indeed seem juvenile for "great" men to exult in childish wordplay, but these puns offer the perfect symbol of what (we *think*) their creators were trying to achieve. They reflect not just a contest of meanings but a contest of cultures and class systems, in which the "heroic" champions of high-minded Western values (represented by a Latin education and a respect for the European classics) were seen to triumph over the peoples, the cities, and the much older traditions of the East. In truth, none of these men may have sent the telegrams for which they are remembered, but our willingness to keep the stories alive says a great deal about why we use language creatively and how we use technologies to communicate.

**Welcome to the Metalevel**

We cherish the few examples of true creativity that survive, in fact or legend, from the bygone age of telegraphy, but Twitter, the modern inheritor of the best aspects of the telegraph, offers us this creativity in freeflowing abundance. Indeed, while the inner workings of the telegraph (or the "Victorian Internet," as writer Tom Standage calls it) had a significant human component, Twitter's end-to-end automation means that our machines are just as capable of sending and receiving tweets as we are.[11] Twitter's application program interface (API) is specifically designed to allow other pieces of software, such as smartphone apps, to exploit all of its *read* and *write* services. These other apps may simply offer intermediary services to their human users, or they may be autonomous consumers and generators of content in their own right. On Twitter we call these mechanical generators of content "Twitterbots," for (ro)bots that have been designed to operate their own Twitter accounts. But are these *bots* capable of the same kinds of educated insolence that we see humans produce on Twitter? Are they capable of generating messages with the same double-edged cleverness and elegant concision as Winkworth's *"Peccavi"* or Benchley's "Streets filled with water. Please advise"? The answer to each of these questions is a qualified yes.

Human creativity is a constantly replenished resource on Twitter, where a stream of newly minted hashtags marks the birth of new challenges to received wisdom and encourages fresh perspectives on the stale and too familiar. Anyone can join in the fun simply by marking one's own tweets with the hashtag du jour or by inventing a new hashtag of one's own to elicit conceptual and linguistic innovation from others. Consider the hashtag *#JamesEllroyStarWars*, minted by the comedian Patton Oswalt to encourage his followers to blend the innocent, fairy-tale world of *Star Wars* with the noirish, argot-heavy world of writer James Ellroy's Los Angeles, a world in which everyone is on the make, on the take, and quick with the slang. This specific tag, one of many minted on Twitter every day, allows Twitter users to show off their knowledge of two very different milieus, yielding such gems as "Ackbar smelled like a plate of calamari, but those bug eyes saw the invisible inevitability. Trap" (from Twitter user *@PearlRiverFlow*) and "Leia kissed Luke on the mouth. Deep down she knew he was her brother, but she grooved on it" (from *@The_Jump_Room*). Twitterbot designers do not set out to replace or supplant this creativity; they simply aim to augment it with what they know and do best: clever engineering. Whether you are a regular Twitter user responding to the creative challenge of how to add your own voice to the game of *#JamesEllroyStarWars* or a bot designer responding to the engineering challenge of building a robot to generate responses that you could never write yourself, the challenge is much the same, if only taken to the metalevel in the latter instance.

There is no reason, in principle at least, why we cannot give our machines enough knowledge of the world to appear educated, or enough metaknowledge to use this education in insolent and entertaining ways. The profound questions of whether a machine can ever feel pride at showing off its "education" to others, or feel surprise at the effectiveness of its own defiance of convention, or feel the joy that comes from being playfully insolent to others are questions we must leave to the philosophers. We can suggest answers in how we go about building and then critiquing our bots, but rather than present a definitive philosophical position, this book instead focuses on ideas, methods, tools, and resources for crafting novel Twitterbots as a well-matched marriage of software engineering and knowledge engineering. Whether our Twitterbots are genuinely creative in their own right, or merely showcases for the metacreativity of their creators, is a question we leave to our readers to answer for themselves. But however one views their efforts, these bots make Twitter a more creative place for everyone.

Most Twitterbots are simple software constructs that make a virtue of their simplicity. Their value is to be measured in ideas rather than in lines of code. Prolific Twitterbot builder Darius Kazemi calls such bots *tiny subversions*, simple creations that amuse and provoke, and whose very artificiality prompts us to think a little more about the nature of human creativity.[12] Kazemi's bot *@twoheadlines* subverts the daily news, creating one imaginary headline from two real ones. Fabricated headlines, such as "Miss Universe attacks northeast Nigerian city; dozens killed," make us laugh and make us think about the baggage we bring with us to a news story (e.g., that stories about Miss Universe are fluff, while stories about Boko Haram are bleak). *@pentametron*, a Twitterbot by developer Ranjit Bhatnagar, finds accidental poetry in the random musings of the Twittersphere. Its technique is simple: find two tweets of ten syllables each that can be read as though written in the poetic meter of iambic pentameter, where every second syllable is stressed and follows an unstressed syllable. Shakespeare's classic line "But, <u>soft</u>! What <u>light</u> through <u>yon</u>der <u>win</u>dow <u>breaks</u>?" is the model of iambic pentameter. By pairing two such tweets randomly if they rhyme on their last syllables, *@pentametron* creates rhyming couplets by pairing tweets such as, "Come on and slam, and welcome to the jam," with, "Many Twitter profiles are filled with spam." Although *@pentametron* lacks understanding of what each tweet might actually mean, its resulting blends are often charming, surprising, and highly retweetable. The bot encourages us to understand each tweet in a new light, and perhaps think a little more, and a little more deeply, about what it is that makes any text worthy of the label "poem."

It takes knowledge—of Twitter, of language, of poetry, of the news—for a bot to insolently spin off new tweets (with new meanings) from the thoughts and words of others. Yet in the popular imagination, the word *bot* is associated with an altogether darker strain of educated insolence, one in which malevolent software agents exploit an inbuilt knowledge of network protocols and security conventions to disrupt and pervert the operation of other software systems. Though the "hackers" who build these systems are not lacking in creativity, theirs is an unbalanced creativity that places an undue emphasis on insolence over education. However, this book is born of the belief that not all hackers are devious and not all bots are insidious. The word "hacker" has an older sense, one that denotes any programmer who takes joy in the pure act of software creation, while "bot" can mean any autonomous software system that is designed to help, amuse, provoke, and even inspire. Although these meanings are not currently the dominant senses of the words hacker and bot, the rise of the Twitterbots in the world of social media is actively reshaping our expectations of software that is both intelligent and creative.[13] This book focuses on this altogether more satisfying and benign, if benignly insolent, world of bot hacking.

### Springtime for Twitter and Irony

Western Union, the most iconic of telegraph companies, sent its last telegram on January 27, 2006, though telegrams had already been viewed as anachronisms for decades.[14] The British post office discontinued the service in 1982, yet the idea, if not the reality, of the telegram still held a secure place in the language and the popular imagination. So, for instance, the ritual of reading "telegrams" from absentee guests at wedding banquets continued unabated, even if the telegrams of old were now replaced with faxes, emails, and texts. It took two months after Western Union killed the telegram for Twitter to post its first public tweet, on March 21, 2006, long before the word *tweet* was even coined (the earliest tweets were called *twitters* or *status updates*). That first tweet, from Twitter cofounder Jack Dorsey, had none of the grandeur of Morse's "What God hath wrought" and showed instead a mix of humility and brand uncertainty when he claimed to be setting up "my twttr." It was, then, more of a small step than a giant leap. But Dorsey's tweet was marking time in more ways than one. Unlike the ephemeral telegrams of yore, the most important instances of which are now found only in the journalistic record (in the best cases) or in apocryphal legend (in the worst, and most likely, cases), this first tweet for Twitter still exists as part of Dorsey's official Twitter timeline. Indeed, the tweet, which has a unique status ID, also occurs in the timelines of many other Twitter users, since it has been retweeted (which is to say, forwarded from user to user) over 100,000 times since its minting. Dorsey's tweet has a URL that reveals its unique status id:

*https://twitter.com/jack/status/20*

Only the earliest tweets have such an impressively low-status ID. Those first tweets were called *status updates* because Twitter was originally conceived as a service that would allow its users to keep friends, family, and other "followers" up-to-date on their comings and goings, that is, on their current status.[15] Dorsey had been inspired by a secondary feature of instant messaging apps that allowed users to explain, with a small piece of text of the "gone fishing" variety, why they were currently unavailable to respond to incoming messages. This status bar was often wittier and more interesting than the actual messages themselves, and it seemed to be a valuable secondary channel for communication in its own right. With colleagues Biz Stone and Evan Williams, Dorsey set out to create an app for mobile phones that flipped this state of affairs upside down: the status message would now become the primary channel of communication between users. As the joke went in those early days, Twitter was an application that allowed you to tell the world what you had for breakfast. Because the service was designed to piggyback on the texting facilities of cell phones, the size of each status update was necessarily limited by the maximum length of a text message: 160 characters. A portion of this 160-character maximum was reserved for use by the app itself, to contain the name (or *handle*) of the sender, a colon, and a space, allowing users to use whatever was left over for their own text. Observing that this practice was unfair to users with long Twitter handles, Dorsey and Stone later standardized the division of space: the app would take just 20 characters for itself and give the remaining 140 to its users. A magic number was born, making 140 a new benchmark for verbal concision. Even as Twitter tinkers with its winning formula and offers users a heftier 280-character container for their thoughts, the platform's original magic number reasserts itself in Twitter's two-for-one arithmetic

The word *tweet* was coined not by Twitter itself but by its users, and was given the official imprimatur of the company only after it gained widespread use. The */status/* in the URL of each tweet is a fossilized reminder of those early days. Showing just how fast time flies, the URL of @*CIA*'s first tweet exposes this ID:

*https://twitter.com/CIA/status/***474971393852182528**

The CIA's first foray onto a social network that promotes both accountability and transparency—"We can neither confirm nor deny that this is our first tweet"—oozes with educated insolence, offering the world a winked acknowledgment of its own lack of accountability and transparency. This sharing of an open secret —*trust us when we say we are not to be trusted*—gives the tweet a self-referential irony that has since caused it to be retweeted over 300,000 times. We know exactly how many times this tweet and others have been retweeted because Twitter itself tracks these numbers, displays them in its app, and makes them available through its API. Though retweeting seems a marquee feature of the service, it was not built in from the beginning. Rather, just as the Big Mac and the Egg McMuffin were invented not at McDonald's headquarters but by individual each one individually. Besides, the notion that a Twitterbot might have God as a follower would be just too good to pass up.

In fact, the operation of this hypothetical *@everynameofgod* is not so very different from a popular and very real Twitterbot named *@everycolorbot*. This bot does exactly what its name implies, at least within the limited world of colors defined by the RGB color standard. Red, green, and blue are additive primary colors, which means that white light can be formed from the balanced addition of all three. Conversely, black is the total absence of all three. All other colors lying between these two extremes can be encoded as a trio of numbers: one for the Red component, one for the Green, and one for the Blue. Using a byte to code for each color component allows us 256 values for each, so three bytes together can differentiate 16,777,216 color combinations. Alphabetically, we can encode each byte with a two-character hexadecimal sequence, giving us sixteen choices (hex (6) + decimal (10) = 16) for each character position. The six-character D8B827 is thus the RGB code for yellow ochre, where R = D8, G = B8, and G = 27. Generating a six-character hex code with this alphabet is no harder than generating a nine-letter name of God from the alphabet of Clarke's monks; indeed, since the names of God must obey certain rules (e.g., no repeat sequences of three letters), we can view *@everycolorbot* as a simplified instance of the same general process. But *@everycolorbot*'s RGB codes have affordances of their own. Each code can also be used to generate a swatch of the corresponding color (a block of yellow ochre, say) that the bot tweets alongside the abstract hex code. *@everycolorbot* currently has more than 130,000 followers, who retweet particular codes and color swatches because they say something about their own aesthetic preferences. Though many believe the world to have started with the words "Let there be light," none have yet formed a doomsday cult that believes the world will end when *@everycolorbot* has tweeted every last one of light's many possible RGB values.

Clarke's story ends not with a loud bang but with an ironic whimper. As the programmers' task nears completion, they worry that the monks will become violent when their religious beliefs are falsified by the world's refusal to end on schedule. They make their excuses and leave early so as to be far away when their program—essentially nine nested for loops— terminates. They trek back down the mountain on ponies, laughing at the monks' strange mix of superstition and technological savvy. After all, the monks have shown more faith in the value of computer-generated texts than they themselves would ever possess. But nearing the end of their trek, Clarke tells us of their alarming observation: "Overhead, without any fuss, the stars were going out." Though the outputs of the program held no special meaning or relevance for the programmers who built it, those outputs found their audience and made their mark, both individually and collectively. A large number of bots, like @*everycolorbot*, work on much the same principle, albeit without the same world-shattering consequences. These Twitterbots are eager generators of tweets that they themselves can never understand, but somebody does, and for those who follow them, that is enough. So if it does not pay to be a naive monk when it comes to automated generation, neither does it pay to be a cynical engineer. Between these two extremes lies a happy medium that the best Twitterbot designers strive to find.

The more we and our Twitterbots generate just because we can, the more our intended meanings get lost in the noise of mere possibility. This point was made dramatically by another short story writer, Jorge Luis Borges, in his tale "The Library of Babel."[20] Borges imagined a vast library of interconnecting rooms whose shelves store every book imaginable. More formally, each book contains 410 pages of 40 lines per page, and each line comprises 80 characters, drawn from an alphabet of 22 letters, a comma, a space, and a period. Within these generous limits, Borges's library contains every book ever written, or a translation of such, as well as every book that ever will, or could, be written. To pick a book at random here is no different from generating one at random, by rolling an alphabetic 25-sided die $410 \times 40 \times 80$ times. So to find any meaning at all in this library, we desperately need a catalog to tell us which books are worth reading and which, by implication, are nonsense. Borges notes that the Library of Babel must inevitably contain such a catalog, insofar as any catalog will itself be just another book that lies within the generative reach of the library. Yet because the catalog is itself a book in the library, it will be lost in a sea of noise and misinformation just like every other book of interest. Indeed, there will be very many catalogs, each claiming to be authoritative, but we can have no idea how to tell those apart without a metacatalog, and so on and on, ad infinitum. It is no wonder that the librarians of Babel are prey to suicidal thoughts, while few readers ever find what they're looking for in this library that has everything.

Borges provides us with some tantalizing examples of the library's contents:

> Everything would be in its blind volumes. Everything: the detailed history of the future, Aeschylus' *The Egyptians*, the exact number of times that the waters of the Ganges have reflected the flight of a falcon, the secret and true nature of Rome, the encyclopedia Novalis would have constructed, my dreams and

half-dreams at dawn on August 14, 1934, the proof of Pierre Fermat's theorem, the unwritten chapters of *Edwin Drood*, those same chapters translated into the language spoken by the Garamantes, the paradoxes Berkeley invented concerning Time but didn't publish, Urizen's books of iron, the premature epiphanies of Stephen Dedalus, which would be meaningless before a cycle of a thousand years, the Gnostic Gospel of Basilides, the song the sirens sang, the complete catalog of the Library, the proof of the inaccuracy of that catalog.

Yet the idea of the Library of Babel is more interesting than the real thing could ever be, just as Borges's selective description of its esoteric highlights is more interesting than the library itself could be, just as a selective summary of the outputs of any wildly overgenerating Twitterbot will always be more interesting than the Twitterbot itself. The Library of Babel is tantalizing because it contains every true answer to every question, every evocative metaphor, every hilarious joke, every stirring speech, every moving elegy, every quotable poem, and every high-impact tweet. But it also contains every wrong answer, every bad joke, every doggerel poem, and every imaginable piece of linguistic excrescence. Without a means to distinguish them all, they are all equally worthless. Borges's library reminds us that creativity is not just about generating the good stuff; it is just as much about not generating the bad stuff. Even the best human creators will produce good and bad in their careers, for a career without missteps is a career without creative risk taking. Naturally, our Twitterbots will also generate a mix of good and bad, of retweetable gems and forgettable dross. Our goal as metacreative bot builders is to achieve a balance between these two extremes.

### Have You Met the Sphinx?

If less is often more in linguistic creativity, Borges tells us that more is almost always less. The engineers of Clarke's story scoff at Buddhist creation myths, but they also lack faith in the power of mere generation alone to achieve meaningful results. Though Clarke turns the tables on his protagonists (and us) to achieve a satirical effect, his surprise ending derives its power to surprise from our shared presumption that generation alone is the lesser part of creativity, just as creative writing is more than putting words and characters on paper. As Truman Capote once said of Jack Kerouac on hearing of the latter's frenetic, Benzedrine-fueled stream-of-consciousness writing method, "That's not writing, that's typing."[21] Our Twitterbots can certainly type, but can they really write? The difference between mere generation—the generation of outputs just because we can, with no consideration of their meaning—and true creativity is easy to see in the extreme cases of "The Library of Babel" and "The Nine Billion Names of God." But this call is much harder to make in the majority of cases, especially when dealing with the outputs of a successful writer or a sophisticated computer. In truth, because sustained innovation is hard and because it is tiring, all human creativity is a mix of unthinking generation and deliberate originality. This makes external critics so crucial to the creative process, because not even the creators themselves can always tell one from the other. When filming *Star Wars* for creator George Lucas, the actor Harrison Ford had a Capote moment of his own, though he expressed his opinion of Lucas's script with less of Capote's signature tartness and more of his own characteristic frankness: "George, you can type this shit but you can't *say* it."[22] If this was Ford's reaction to Lucas's first *Star Wars*, we can only imagine his views of the overstuffed, more-is-less excess of the three prequels.

Lucas raided the kitchen cupboard of pop culture ideals and narrative tropes when he made those first *Star Wars* movies. Yet even with its leaden exposition and its corny dialogue, Lucas managed to plant the seed of something of lasting value. But not all critics took this benign view. In her book *When the Lights Go Down*, Pauline Kael had this to say about the film: "It's an assemblage of spare parts. ... *Star Wars* may be the only movie in which the first time around the surprises are reassuring. ... The picture is synthesized from the mythology of serials and old comic books."[23] With *American Graffiti, Star Wars, Raiders of the Lost Ark*, and *Willow*, Lucas had turned nostalgia for lost innocence into an identifiable shtick. So it's not surprising that the futuristic world of *Star Wars* is set A Long Time Ago (or that the villain of *Willow* is named *General Kael*). When you see this approach once, in a film like *Star Wars* or *Raiders*, it seems charming and fresh, if not very original: it really works! When you see it in film after film, it becomes a gimmick and begins to resemble mere generation more than true creativity. It takes effort to avoid repeating oneself and self-knowledge to recognize when one has. Even the greats, like Picasso, occasionally lapse into lazy self-pastiche, finding themselves unthinkingly doing the same things and repeating the same patterns over and over. When discussing a painting he considered one of his lesser works, Picasso memorably dismissed it as a fake. When pressed on the matter—for the indignant owner claimed to have seen Picasso work on that very picture in his studio—Picasso is said to have shrugged and said, "So what? I often paint fakes."[24] But unlike master artists such as Picasso, most bots are defined by their shtick and cannot easily transcend it. See enough of these bots' outputs, and we see all their gimmicks laid bare. This is not to say that gimmicks are always a bad thing; rather, any particular gimmick should be used sparingly, perhaps in unpredictable combinations with others, and with enough self-knowledge to know when it is time to put an overused gimmick back on the shelf for a spell.

Like humans, Twitterbots work at various scales of complexity and ambition. Ambitious designers aspire for their best bots to operate as thought-provoking conceptual artists, exploring a space of often surprising possibilities, while others are built to be the software equivalent of street performers, each plying the same gimmicks on the same streets to an ever-changing parade of passersby each day.[25] While the locals may soon tire of the same old shtick—the bot equivalent of the moving statue or the levitating man—each day brings new faces with new smiles and the occasional round of applause for the same reassuring surprises. A bot like *@EnjoyTheMovie* is clearly designed to deliver its share of familiar surprises, by tweeting spoilers to random Twitter users who unwisely express an interest in seeing a movie with a well-guarded twist. Tweet that the corn is a-poppin' for an evening in front of the box to watch *The Sixth Sense*, and *@EnjoyTheMovie* will joyously ruin the movie by revealing that Bruce Willis is dead all along. Or tweet even a passing interest in seeing *The Crying Game* and the bot will spoil the midmovie transgender twist. Or at least the bot *would*, if it were not the target of sustained reports of abuse from its many victims—the kind of reports that get a bot suspended on Twitter. But this is very much the point of *@EnjoyTheMovie*. The bot is designed to ply its fixed repertoire of familiar surprises in ways that provoke the ire of its targets, and it is this ire—expressed with often hilarious profanity—that yields the truest and most affecting variety. This bot is the kind of street performer who makes sport of some unlucky tourists to earn laughter and applause from others.

But we should not be overly critical of Twitterbots with a limited repertoire that are designed to do just one kind of thing, especially if they do that thing well and to our amusement. Cocktail parties and country clubs are full of humans who operate in much the same way, telling the same old jokes, performing the same old tricks, using the same old catchphrases ("that's what *she* said!"), and dining out on the same old anecdotes that grow with the retelling. These people live in a temporal-distortion bubble where no gimmick ever grows old. It is our lot to live outside that bubble, if only to burst it occasionally with a pinprick of reality. Consider the following exchange from the 1999 comedy *Mystery Men*, a film that follows the misadventures of a group of wannabe superheroes with rather underwhelming powers.[26] Mr. Furious has anger issues, while the Sphinx's only power is an ability to torture syntax until it yields a phony profundity:

**The Sphinx:** He who questions training, only trains himself in asking questions. ... Ah yes, work well on your new costumes my friends, for when you care for what is outside, what is inside cares for you. ... Patience, my son. To summon your power for the conflict to come, you must first have power over that which conflicts you.

**Mr. Furious:** Okay, am I the only one who finds these sayings just a little bit formulaic? "If you want to push something down, you have to pull it up. If you want to go left, you have to go right." It's ...

**The Sphinx:** Your temper is very quick, my friend. But until you learn to master your rage ...

**Mr. Furious:** Your rage will become your master? That's what you were going to say. Right? Right?

**The Sphinx:** Not necessarily.

Or rather, "Yes, necessarily," for the Sphinx has hit on a successful gimmick for mere generation that turns casual utterances into guru-like prognostications. His shtick can appear deep, yet his rhetorical strategy is little more than repetition with crossover. The strategy is an old one that has been studied since antiquity under the name *chiasmus* (where *chi*, the cross-shaped Greek letter $\chi$, signifies crossover). One may imbue the Sphinx's utterances with real meaning, perhaps even a profound truth, but it seems clear that for this *professor $\chi$*, meaning takes a backseat to surface form in his drive to seem wise and all knowing. Despite his short fuse, Mr. Furious has seen enough for his critique to be on target. The Sphinx could no more change his strategy if he were a Twitterbot, @*ChiasmusBot* say, forever fixed in its programming to perform the same trick over and over.

We meet people like the Sphinx all the time at social gatherings where a glib affability is encouraged, such as—sadly—academic cocktail parties. Indeed, the term *cocktail party syndrome* was coined to describe just this kind of sonorous chatterbox, always ready with a glib humorous response or an affable blend of clichés and platitudes. But cocktail party syndrome (aka *chatterbox syndrome*) is also used, more formally, by clinical psychologists to label the cluster of sociolinguistic traits that are often observed in children affected by hydrocephalus.[27] The children who present with this syndrome are extremely loquacious and may appear highly sociable, yet this apparent verbal acumen conceals impaired social skills and a lower intelligence overall. Chatterbox syndrome allows these kids to speak with confidence and apparent knowledge of topics they know little about, using words whose true meanings are lost on them. Because the child's knowledge of words and the ways they chunk into larger syntactic units exceeds his or her understanding of the meaning of those units in context, a child with chatterbox syndrome can sound and act remarkably like a Twitterbot. As we saw with the Sphinx, a fictional creation that distills the traits of the many phonies we have all met into a single caricature, adults occasionally exhibit the same traits without a clinical diagnosis to excuse them. Chatterbox traits are simply more pronounced in certain children (or in certain software systems) who use words to impress rather than to communicate. In the following excerpt from a 1974 study by Ellen Schwartz, the word *child* might well be replaced with *Twitterbot* and the word *he* by *it* without affecting the validity of its core message: "The *child* uses automatic phrases and clichés; at times *he* even quotes directly from television commercials or slang *he* has heard others use. *He* uses words from other contexts that almost but not quite fit his conversation." Both *@pentametron* and *@twoheadlines*, two bots that assemble their tweets by directly quoting from others, benefit from the slightly discordant note that emerges when two quotations that almost but not quite fit are forced together. Note how Schwartz's description of a child with chatterbox syndrome is also strikingly similar to Pauline Kael's description of George Lucas's patchwork creation, *Star Wars*. The chatterbox child treats language as its own source of cultural spare parts that can be recombined into an assemblage of familiar surprises: *familiar* because the parts are each so familiar, *surprises* because they may be put to jarring new uses that are not at all what we expect. So while the generative creativity we see in Twitterbots may be artificial, it sits on the very same continuum of magpie bricolage that links the chatterbox child to the adult Sphinx to the successful maker of blockbuster movies. It is a continuum we shall explore extensively with the aid of Twitterbots in this book.

## Race You to the Bottom

In a study from 1983, the sociologist Neil McKeganey offers this assessment of a child, "Linda," with chatterbox syndrome: "Linda lives in language and loves to talk and listen to it. She does not however always grasp the meaning and is inclined to indulge in the sound and play of words."[28] Like Linda, most Twitterbots "live in language" rather than in the real world. Their knowledge, such as it is, concerns words and their norms of association, and not the world in which their followers live. Although conversations with children like Linda can be engaging but disorienting for adults, these kids show an obvious love of language that bot builders often strive to capture with their own creations, with similarly jarring results. Yet whether we are dealing with a Twitterbot or a chatterbox child, the sophistication of their language can lead to unreasonable expectations that may, respectively, cause disillusionment for the bot's followers and frustration for the child. McKeganey quotes a doctor who says this of Linda: "Talks like a grown up Yankee. Incredibly charming. Incredibly vulnerable. Adult language, infantile frustration threshold." Our charming bots may not have Linda's vulnerability to slights, but in their own way, they can be just as fragile and lacking in robustness. How Twitterbot designers react to this fragility will decide whether their bots are designed to openly engage with, or merely deceive, their human followers.

Though Linda's reach for words far exceeds her grasp of their meaning, hers is not a pretentious use of language and its possibilities. For unlike phonies like the Sphinx, kids like Linda live *in* language rather than *behind* language, and so their loquaciousness is born more of exuberance than of deceit. The question for us as bot builders is whether we are driven more by the latter than the former, to build systems that aim to keep users from the truth rather than invite them in. If the public discourse about artificial intelligence has it that AI is primarily about the building of fake humans that can pass for real people in online dialogues, then AI is as much to blame for this corruption of its ideals as the popular press. From its roots in the modern era, when Alan Turing first proposed what is now called the Turing test in 1950, AI has been seen as an imitation game.[29] Turing's idea for a language-mediated test of intelligence and humanity has since become a science-fiction staple, but for Turing, this test was merely a thought experiment with which he hoped to peel away the veils of cultural and spiritual prejudices that we humans naturally bring to any consideration of nonhuman intelligence. If we can have a probing conversation with another agent about our feelings, our ambitious, our hobbies, our passions, our favorite movies and books, and not be able to tell whether we are speaking with another human or a machine, then that other agent must surely possess a level of intelligence that is, for the purposes of everyday conversation at least, just as real as a human being's. Like a foreign sleeper agent with the deepest of deep covers, to fake it this well requires a machine to truly become what it is pretending to be. Consider this excerpt from Turing's 1950 paper, where he imagines a human interrogator interviewing a "witness," a writer of sonnets who may or may not be another human being:

**Interrogator:** In the first line of your sonnet which reads "Shall I compare thee to a summer's day," would not "a spring day" do as well or better?
**Witness:** It wouldn't scan.

**Interrogator:** How about "a winter's day?" That would scan all right.

**Witness:** Yes, but nobody wants to be compared to a winter's day.

**Interrogator:** Would you say Mr. Pickwick reminded you of Christmas?

**Witness:** In a way.

**Interrogator:** Yet Christmas is a winter's day, and I do not think Mr. Pickwick would mind the comparison.

**Witness:** I don't think you're serious. By a winter's day one means a typical winter's day, rather than a special one like Christmas.

Notice that Turing does not imagine his interrogator talking to a child-like Linda or to an adult like the Sphinx. He imagines a free-flowing dialogue between two well-educated adults, with an awareness of the classics and who can speak with ease about their feelings and impressions, or about complex cultural events such as Christmas. Indeed, Turing compares his test to a viva voce examination of a PhD candidate, in which an expert examiner interviews—though *interrogates* really is a more apt verb—a student about their research topic to see if they truly understand that topic in depth. Nobody passes a "viva" like this by subjecting the examiner's words to a chiastic hernia or by wandering off topic and randomly quoting TV advertisements, jingles, and slogans. Yet the latter approach is not so very far from the mark when it comes to modern approaches to the Turing test. Since we still lack a sufficiently robust, knowledge-based technology to allow a machine to interact with a human with the deftness of the "witness" in the dialogue, the Turing test has instead been debased to the point that it resembles not so much *The X Factor* or *America's Got Talent* but *The Gong Show*. While the purpose of a viva voce examination is not to fool but to impress an examiner, the same can no longer be said of Turing's Test in its modern guise. It has become instead a faker's charter.

Though we cannot expect our bots to interact with humans in the same way as Turing's imaginary witness, this is not what Twitterbots have ever been about. Twitterbots are not fake humans, nor are they designed to fool other humans. Yes, it can be satisfying to see a passing Twitter user retweet or favorite the wholly fabricated output of one of our bots, in the belief that human intelligence was responsible for both its form and its meaning. We might even consider this eventuality—not a rare occurrence on Twitter, by any means—as yet another successful instance of a 140-character Turing test. But the truth is much simpler and just a little stranger: humans do not follow Twitterbots because they believe them to be human. Humans follow bots because they know them to be artificial and appreciate them all the more for this otherworldly artificiality. Every Twitterbot, no matter how simple or sophisticated, is a thought experiment given a digital form.

Though many bots are one-trick ponies like the Sphinx, each embodies a hypothesis about the nature of meaning making and creativity that encourages its followers to become willing test subjects. Naturally, bot designers want to impress their followers as clever metacreators, but they also encourage these followers to speculate on the workings of their Twitterbots and notice when additional sophistication and new features have been added. Because bots are artificial but use human language and other systems of human signification to speak to human concerns, Twitterbots blur the line between the natural and the artificial. They show us how human meaning can arise via nonhuman means and reveal the hidden regularities at the heart of human behavior. So when we humans interfere with the autonomy of a Twitterbot, so that its outputs result from *artificial* artificial intelligence rather than wholly mechanical means, the bot's followers naturally feel cheated and betrayed. The scandalous case of *@horse_ebooks*, a once-popular Twitterbot whose case we discuss in the next chapter, shows how Twitterbots turn the logic of the Turing test on its head: it is the possibility of humans pretending to be machines, not machines pretending to be humans, that most exercises those who build and follow bots on Twitter.

**¡Viva la Revolución!**

George Orwell said that every joke is a tiny revolution, a tiny attack on the facade of received wisdom that suggests the whole edifice is riven with fault lines.[30] Orwell's take on jokes is echoed in the words of bot builder Darius Kazemi, who builds his Twitterbots to be tiny subversions of the status quo. But whether we view Twitterbots as thought experiments, jokes, tiny subversions, or even tiny revolutions given digital form, these bots are typically small, idea-driven systems about which it is rather easy to make such big claims. We can talk big and at length about such systems, even when their behaviors can be captured in the simplest of rules and coded in the shortest of programs. Yet this is true of many domains of creative endeavor, for creativity is the ultimate cognitive lever. With the application of a shrewd insight at just the right time and in just the right place, a modicum of productive novelty allows us to reap disproportionate yields with surprisingly little effort. Twitterbots are just one more domain of endeavor in which the lever of creativity allows us to turn less—less code, less effort, less restraint—into more—more outputs, more diversity, more surprises. In that spirit of less is more, it is time to stop talking about Twitterbots in the general and start talking about them in the specific, and indeed, to start building these autonomous generative systems for ourselves.

In the chapters to come, we start at the very beginning, by registering the app that will become our first bot. We consider the role of the Twitter API and how it can be accessed by our own software systems—via a third-party library such as Twitter4J—to do all of the things that we humans do on Twitter. Just as chefs joke that beef tastes so much better when cooked in butter because these ingredients grow up together in the same cow, we conduct our exploration of Twitter and Twitterbots through Java, a popular programming language that has grown up hand in hand with the Web and contributed so much to the Web's success. We can use other languages to write our Twitterbots, of course, with Python being another popular choice. However, our intent in this book is to focus more on the ideas and the principles that drive our bots than the specifics of their code. So when it more convenient to do so, we shall sketch only the broad strokes of the code, pushing the specific details into the website that accompanies this book. Though it is important to show real Twitterbots in the flesh, we do not want the code to get in the way of a real understanding of what that code is designed to achieve. So our promise to nontechnical readers is this: you don't have to understand any of the code to understand the point that is being made, as our arguments will never hinge on the peculiarities of any programming language or platform. We take our cue in this regard not just from Alan Turing but from Ada Lovelace, the nineteenth-century mathematician who is now deservedly celebrated as the first "programmer" in the modern sense of the word, despite never having had a physical computer to program. Countess Lovelace is a singular figure in the history of computing, having succeeded at uniting the poetic tradition of her father, Lord Byron, with the scientific tradition of inventors such as her mentor, Charles Babbage, to found a whole new tradition of her own. This bridging of frequently antagonistic traditions was dubbed "poetical science" by Lovelace herself, and few other names seem quite so suited to the modus operandi of modern bot designers.[31] As we'll see throughout this book, even when occasionally flirting with code, Twitterbots are much more about ideas than they are about method calls, and they have as much to do with poetry and art as they do with science and engineering.

## The Best of Bot Worlds

If Twitterbots were magic tricks, we hope this would be the kind of book that would get us drummed out of the Magic Circle. Twitterbots are not magic tricks, of course, even if they share some obvious similarities: each exploits the foibles of human psychology to amuse and surprise, and viewing each can appear mystifying at first. But whereas tricks involve deception and concealment—hence, the Magic Circle's warning to any magician who dares to reveal the workings of a trick to the public— Twitterbots are designed to be open, about their artificiality and their inner mechanics. When a practitioner such as the infamous Masked Magician (of the Fox TV specials) pulls back the curtain on how the big Las Vegas– style tricks are really performed, the viewer can feel cheated because the sum of the mechanics often adds up to much less than the value of the illusion. In such cases, more definitely produces less, especially since a cultural cynicism about smoke and mirrors has long since inflected our response to stage magic. However, such practitioners have a larger and more laudable goal than winning audience share. By dispelling the mystery around the tried-and-true favorites of the cabaret hack, such exposés spur other magicians to invent new and more creative ways to renew their hold on the audience's imagination. As Twitterbot designers rarely conceal the workings of their systems—indeed, concealment is often impossible, since most bots wear their generative principles on their sleeves—there is no comparable sense of *gotcha!* and no hacks to brusquely push aside. For where a magician says, "Look ye and wonder at the mystery of my magic," the bot designer says, "Look. It really is this simple, so go do it for yourself." While magic exposés produce stifled yawns, lifting the curtain on a clever bot can be a most satisfying experience, not because it shows how easily we can be fooled—since so few bots pretend to be human—but how easy it is to be creative if we put our minds to it.

With this book, we want readers to experience the best of bot worlds. It would be disingenuous of us to ask you to pardon this pun, as we intend to force it on you more than once. Our upcoming survey of the world of Twitterbots is called (you guessed it) *The Best of Bot Worlds,* and the website that accompanies this book (a trove of tools and data for building your own creative bots) can be found at http://www.bestofbotworlds.com. So, drum roll please, and on with the show.

## Trace Elements

You will by now have noticed that our preferred spelling for "Twitterbot" reflects a pair of choices regarding capitalization and spacing that are far from universal in online discussions of these magical little programs. Because *Twitter* is a proper noun, it seems natural to capitalize its uses in text, and we follow this pattern in also choosing to capitalize the word *Twitterbot.* Perhaps it is the informal and often playfully subversive nature of bots that leads many online to describe them as *twitterbots* with a small *t*, but in truth the decision to capitalize or not carries very little meaning. We do it here for consistency if for no other reason. While online discussions are just as likely to insert a space between "Twitter" (or "twitter") and "bot," we also show a preference here for the solid compound *Twitterbot* over either *Twitter bot* or *twitter bot*. Though bots come in many varieties and can operate across a diversity of platforms, our focus here sits resolutely on the bots that operate on Twitter and nowhere else. This book explores how Twitterbots exploit the unique affordances of Twitter to squeeze an extra measure of magic from language and social interaction, and our spelling of *Twitterbot* is intended to signify the special bond between bots and their host.

## 2 The Best of Bot Worlds

### Top of the Bots

Every hour, on the hour, the tower of the Palace of Westminster in London explodes with the sounds of bells ringing as Big Ben strikes out the time in a series of resounding bongs. At the same time, humming away on a server without quite as much applause, a script sends a tweet to *@big_ben_clock*'s Twitter feed, with the word *BONG* typed out one or more times to signify the hour. More than 490,000 people follow the unofficial Big Ben Twitterbot, which has been tweeting the hour since 2009. It even updates its profile picture with images of fireworks every New Year's Eve and little fluttering hearts on Valentine's Day. Moreover, it continued to tweet on the hour even as the real thing went silent in August 2017 for a projected four-year period of rest and restoration.

Getting a computer to post tweets for you goes back much further than 2009, however. In chapter 1, we recalled Twitter's very first tweet, "just setting up my twttr," from Twitter cofounder Jack Dorsey. But even that first tweet was sent not with a web interface or a mobile app, but through a script running on Dorsey's computer. A few months after that very first status update, Twitter released the initial version of its application programming interface (API), a special tool kit for interacting with a particular website, technology, or program that exposes all of the public functionalities of a service. Twitter's API would let people write programs that could tweet for themselves, whether it was just a series of bongs every hour on the hour or something much more complicated.

Today, automated Twitter users, or bots, come in all flavors, shapes, and sizes. In fact, in 2017 it was estimated that as much as 15 percent of Twitter's users were not humans but bots. Most of those Twitterbots are of a less-than-edifying variety—little automated advertisers that wander the platform to try to convince users to click on a link or look at a picture.

These advertising bots watch for specific keywords and hashtags to find the right people to target, pester, or poke. But within this enormous, writhing mess of cynicism, we can find little software gems like *@big_ben_clock*, as unique as they are silly, designed to make us smile, frown, or think about something else for a moment. Wandering through this weird world of strange software can feel like blazing a trail through an alien jungle, but just like botanists trying to categorize new discoveries, we too can try to name families of Twitterbots and group them together according to the features and ideas that they have in common. In doing so, we can unpack the monolithic idea of a Twitterbot into different facets, each one a little easier to understand. We'll see that bot builders have diverse reasons for making bots, and a comparable diversity holds for those who follow them too. We'll see that bots can be playful, aggressive, thought provoking, or entirely serendipitous. Along the way, we may even get ideas for Twitterbots that do not yet exist. Our proposed taxonomy for bots in this chapter is just one possible way of classifying this amazing family of software agents. As we explore the different categories, you might forge your own connections between bots, or invent new categories that we fail to mention, or identify bots that fall into multiple categories at once. In reality, every bot is unique, so do not worry if you occasionally disagree with our groupings.

The first and simplest kind of bot is a *Feed* bot. Feeds are bots that tweet out streams of data, usually at regular intervals, and usually forever. Some Feed bots, such as *@big_ben_clock*, tweet out their own kind of data (in this case, bongs according to the current time) in a special arrangement. Other bots tweet out information from large, richly stocked databases. Darius Kazemi's *@museumbot* tweets four times a day, and each tweet contains a photograph of an item from New York's Metropolitan Museum of Art, thanks to the museum's open-access database of its massive collection.[1] Feed bots are simple and elegant, which can make them attractive options for bot builders. One of the most famous Twitterbots ever written, Allison Parrish's *@everyword*, used a dictionary as its database, and it tweeted (during its lifetime) every English word in alphabetical order, two per hour, from start to finish. Today it lies dormant, having exhausted its word list, but at its peak, the bot enjoyed ninety-five thousand followers who hung on its every word. We'll return to the strange cult of *@everyword* later in this chapter.

Feeds can also create new sources of data, as well as recycle data that already exist. A common kind of Feed bot is one that mixes words and phrases together, as selected from bespoke word or phrase lists that are created by the bot builder. *@sandwiches_bot*, now sadly dormant, generated randomized ideas for sandwiches by combining ingredients (like shredded carrots and thinly sliced chicken, or bread types such as focaccia), presentation styles (stuffed with, topped with, garnished with), and a special list of names. So you might well end up with something like this in your lunch-time tweet: "The Escondido: Focaccia stuffed with thinly sliced chicken, brie, red cabbage and watercress topped with spicy mustard." While not every sandwich turns out to be a winner, the Escondido enjoyed two retweets and three favorites from the bot's followers, so perhaps a user actually contemplated making one. Every day at lunchtime, the bot produced a new sandwich, and its followers delighted at seeing what might pop up next. Good or bad, it is very unlikely to be something they will have seen before on a menu.

In 1989, the British comedy duo Stephen Fry and Hugh Laurie performed a sketch they called "Tricky Linguistics" on their TV show, in which Fry mused about the vast scale of the English language and the unique beauty this confers on any sentence.[2] Fry was reveling in an insight that linguist Noam Chomsky had made famous before him, that the raw creativity of human language allows any one of us to invent, on the spot, a seemingly meaningful utterance that no other person has previously uttered or thought in human history.[3] Fry's framing is perhaps more amusing than Chomsky's: "Our language: hundreds of thousands of available words ... so that I can say the following sentence and be utterly sure that nobody has ever said it before in the history of human communication: *'Hold the newsreader's nose squarely, waiter, or friendly milk will countermand my trousers.'*" When we check Twitter at lunchtime and see that *@sandwiches_bot* has created another culinary masterpiece (or not), we get a little taste of what Fry is alluding to here. We get a sense that this combination may never have been seen before, and however slight that revelation might be, it tickles us. Fry's musing also bears some resemblance to the philosophical notion of the sublime, the sense of wonder and awe that is evoked when we come face-to-face with the immensity of nature.[4] Philosophers in the eighteenth century would note the extreme emotions they felt on trips through the Alps when faced with the realization of their insignificance relative to the scale of the universe and of time itself. While *@everyword* and *@sandwiches_bot* cannot compete on this romantic scale, there is an undeniable beauty associated with watching a slow and inexorable process—such as the printing of every word in the dictionary (or of every name of God) – finally come to completion. The word *sex* was retweeted by 2,297 people when *@everyword* finally reached it, and part of the reason (beyond juvenile titillation) must surely be the shared feeling that this was always certain to happen, that these people had witnessed it, and that it was never going to happen again.

This strong connection to expectation, in terms of both the data being tweeted and our ideas about how computers work, can also produce even stronger emotions. When *@everyword* eventually reached the end of its list of words beginning with *z* (with "zymurgy," which earned a strong 816 retweets because of this sense of finality), its followers expected the ride to be over. But an hour later, *@everyword* tweeted a new word: "éclair." Replies to this tweet conveyed both anger and surprise. One follower called it the "GREATEST RUSE OF 2014," another "utter chaos." While they were no doubt playing up their emotions for an audience, the tweet certainly came as a huge surprise. To a speaker of English, *z* is the last letter of the alphabet. But to a computer, accented letters such as *é* have internal codes that numerically place them after the letter *z*. *@everyword* continued for another seven hours before finally coming to a genuine halt.

### Qui Pipiabit Ipsos Pipiodes

Feeds bots are ways to include a new kind of tweet in your feed, whether it's daily recipes for questionable sandwiches or simply the word *BONG* cutting up your feed into hourly chunks. Other kinds of bots do not wait for you to come to them for information, however; rather, they come to you. We call this kind of bot a *Watcher* bot.

Unless your account is protected, meaning its tweets cannot be viewed without your permission, every tweet you dispatch is fired off into the void for everyone to see. Sometimes a void is exactly what Twitter feels like: an empty space where no one replies and your sentiments, no matter how desperate, vanish forever into the ether. This can be a real problem when those sentiments include a cry for help. Several Twitter users noticed this problem and wrote Twitterbots to help, such as *@yourevalued* by *@molly0x57* and *@hugstotherescue* by *@sachawheeler* (now both inactive). *@yourevalued* periodically searched for the phrase "nobody loves me" in tweets, and when it found an instance, the bot replied with one of a number of random responses, including an emoji heart, or the phrase "I like you." The bot's profile picture is a white square overwritten in black with "You Matter." While it's not quite the same as human affection, the bot's responses can often be surprising or even funny to someone who is not expecting them. The bot cannot change someone's world or solve anyone's problems, but for a brief moment, it can intervene in someone's life to remind them that they are valued; they exist and matter, if only because someone (or something) else has taken notice of their tweets. Both *@yourevalued* and *@hugsto-therescue* identified themselves as bots, either in their Twitter names or their profile biographies. This is important because it prevents the bot from posing as a real human and perhaps causing further pain by disappointing someone later on. Nonetheless, neither bot is operational at the time of writing, with *@yourevalued*'s bio citing a conflict with Twitter's terms of service for its indefinite hiatus. This is our first encounter with what some bot authors call *Twitterbot ethics*, a code of conduct for people writing the software that lives on Twitter. We return to the question of ethics repeatedly in this book, including later in in this chapter. Suffice it to say that not all bots play by the rules, as we shall see.

*@yourevalued* quietly replied to the people whose tweets it discovered, but other bots reuse their finds in their own public tweets, in a combination of Feed and Watcher functionality. *@ANAGRAMATRON* by *@cmyr* searches for tweets that are anagrams of one another and retweets them in pairs. Because the tweets are plucked from the public feeds of real users, the results are thus unpredictable and often fascinating. It can bring a smile to your face to realize that "I hope it's not bad man" is indeed an anagram of "time to abandon ship." A bot we met earlier, *@pentametron* by *@ranjit*, also plays with this idea, searching for tweets that can be scanned in iambic pentameter (a poetic meter built around groupings of ten syllables, with the stresses landing on alternating syllables). It retweets these tweets as rhyming couplets that sound perfectly compatible and neat at a poetic level but also possess an unfiltered rawness that emerges from its sourcing of tweets from all corners of Twitter. While *@pentametron* may not see a semantic reason to pair "Sure hope tomorrow really goes my way" with "Just far too many orders to obey," we humans are easily persuaded to find unifying reason behind the superficial rhyme.

Watchers like *@ANAGRAMATRON* and *@pentametron* are, like many other popular bots, entertainingly unpredictable. They also have another interesting quality that goes some way toward explaining their popularity, for they exude a sense that they are bringing order to the messy, chaotic, and enormous world of Twitter. While the philosophical idea of the sublime might encourage us to feel small and powerless against the scale and onslaught of Twitter's digital Alps, Watcher bots organize and sort the millions of tweets being sent every second into neat piles. These ones rhyme. Those ones are anagrams. Even bots like *@yourevalued*, working away in private, are designed to fight back against the roiling flood of tweets, picking people out to remind them that they are not lost and ignored amid all the havoc of social media.

Other Watchers work with different aims in mind. *@StealthMountain* is not the kind of bot that you follow if you want it to notice you. When it needs to and when your behavior warrants it, the bot will find you. It searches for any tweet containing the phrase *"sneak peak"* (as opposed to "sneak peek," meaning an exclusive preview) and publicly asks users if they have made a spelling mistake. It is a remarkably simple bot that repeats the same shtick time after time, but because it finds *you*, to point out *your* mistakes, the bot is not just surprising but sometimes immensely aggravating too (as one user put it, "GO AND JOIN THE GRAMMAR POLICE"). But not everyone minds, and many users reply to thank the bot. Nevertheless, *@StealthMountain* is a good example of a bot that does something that we humans may not feel so comfortable doing for ourselves. Watchers thus stand on somewhat shaky ground when it comes to bot ethics, because Twitter frowns on bots that send unsolicited messages to users who are not also followers. This is largely because unsolicited contact is a key tactic of the spam bots that send click bait, advertisements, or worse to thousands of users each hour. Even benign bots with a positive mission, such as *@yourevalued*, can fall afoul of these restrictions, as Twitter tries to grapple with where to draw the line for acceptable Twitterbot conduct. So *@hugsto-therescue* no longer exists on Twitter, while *@yourevalued* remains shuttered for much the same reason.

### Interaction Hero

The bots we have seen so far are relatively passive, going about their business regardless of what anyone else does. But another kind of bot is the Interactor: responsive bots with special behaviors that are designed to talk back to the users who poke them. One such Interactor is *@wikisext* by *@thricedotted*. *@wikisext*'s main feed is a stream of tweets that resemble the language used in sex texts, or sexts, euphemism-laden and highly suggestive texts sent privately from one person to another. *@wikisext* trawls the how-to pages of a website called wikiHow in its search for sentences that can be twisted into sexualized euphemisms.[5] Even a how-to page about homebrewing might contain promising sentences such as, "Obtain your brewer's license" or "Choose one or more yeast strains." *@wikisext* shifts the pronouns and verb endings to rephrase these in a more suggestive style directed squarely at readers, such as, "I obtain my brewer's license... you choose one or more yeast strains." Because we have been told that its tweets are euphemistic, we can read sexual meanings into the bot's most bizarre non sequiturs, which is where *@wikisext* gains much of its comedic power. We may not know what "you begin by touching my crucifix and praying the sign of the cross" might mean in explicit anatomical terms, but we hardly need to since readers can interpret it and visualize it to whatever degree they desire. *@wikisext* pushes at its limits at times, yet because it never explicitly says anything rude, it can get away with so much. This is exactly why it can be so much fun to follow on Twitter.

*@wikisext* is an Interactor bot because it also replies to the tweets directed to it with a new sext tweet, which is generated in the same way as the stuff of its main feed. While this may seem like a simple feature, it forms a crucial part of the bot's appeal, because it encourages its followers to play with the euphemistic power of language too. Browse *@wikisext*'s tweets, and you can see conversations with the bot that extend over many tweets as replies and counterreplies shoot back and forth. Though the bot has no sense of continuity, so that subsequent tweets tend to be drawn from different how-to pages and topics, the challenge of improvising a response seems to entertain those who engage with it. While the idea of sending sexually suggestive tweets to an impersonal piece of software that reads self-improvement articles might not appeal to everyone, surely some of the aesthetics of the bot movement hinge on this joy of interaction with an unpredictable agent. For many, there is a sense of mystery in how an algorithm might work or in what it might do next. Even for seasoned programmers who can guess at a bot's functionality, there is still a delight in poking the bot and waiting for it to poke back. Whatever will it say next?

Some bot builders take this aesthetic and make it central to a bot's design, and one bot in particular, *@oliviataters* by *@robdubbin*, is famous (and infamous) among bot authors for this very reason. Dubbin's intention was to create a bot that would tweet like a teenage girl. *@oliviataters* tweets about various teenage concerns such as dating, growing up, Taylor Swift, and selfies, as in: "i wonder by this time next year i will have asked for a selfie stick for Christmas. why? why would you?" But the bot does more than just tweet, for like *@wikisext*, it also responds to replies and actively seeks out new user interactions. This can mean "favoriting," or replying to tweets that it likes or starting conversations out of the blue with its followers. While *@oliviataters* may have just under seven thousand followers at present, this rich interaction fosters a large measure of devotion among its fans, who personify the bot and converse with it on a regular basis. When the bot was suspended in May 2015, a small but successful campaign, complete with its own hashtag, *#FreeOlivia*, was launched to get it reinstated. So the Turing test be damned: *@oliviataters* made its followers care about a bot, which is surely one of the most powerful kinds of interaction there is.

Interacting with Twitterbots like *@oliviataters* and getting excited by their personalized replies is not a new affordance of technology. In the 1960s, Joseph Weizenbaum, a computer scientist at MIT, wrote a now infamous piece of software called ELIZA.[6] The software, named for the leading lady in *Pygmalion* and *My Fair Lady*, was designed as an AI experiment that inadvertently became a landmark example of early interactions between humans and computers. Specifically, ELIZA was used to mimic the soothing interactions of a psychotherapist who asks pertinent questions and responds appropriately to the replies of a patient.[7] ELIZA seemed quite convincing to many, drawing in otherwise intelligent humans (such as Weizenbaum's secretary) to reveal their most private concerns. In reality, ELIZA would carefully select from a database of stock responses and lax templates, speaking in a way that frequently deflected the conversation back to the user. Responses such as, "Why do you say that?" and, "Why do you feel that way?" need little context but imply a great deal. ELIZA is a fascinating example of our relationship with software, a relationship that has evolved and become even more complicated since Weizenbaum's day. Even when people were told precisely how ELIZA worked—that is, when they were told that the software had no real understanding of either psychology or their personal situations—they still viewed the system favorably and continued to use it.[8] How we feel about a piece of software, how we personify it, and how much of ourselves we bring to interactions with it are all qualities that affect and strengthen our investment in any given piece of software. This is as true of today's Twitterbots as it was of ELIZA in the 1960s.

We return to this idea at various junctures in this book because our strange, evolving relationship with technology is where Twitterbots sprang from, it is why they survive and flourish on social media, and it remains an integral part of their future. Some of the most exciting and unusual Twitterbot stories, as a result, emerge from Interactor bots.

## Mashed Botato

Bots that tweet, bots that search, and bots that talk back: almost every bot falls into one of these categories. But there are many other labels we can apply to certain kinds of bots, to understand why they are made and what draws certain people to follow them. Instead of broad categories, these labels mark out small subgenres or niches. One populous niche is the mashup: bots that mix together different textual sources. A common mode of mashup is the eBooks-style bot. On Twitter today, the suffix *ebooks* in a user's Twitter handle typically (though not always) signifies that an account is a bot account and has been set up to mimic the user whose name precedes the suffix or sounds similar to it. Thus, for example, *@LaurenInEbooks* is the eBooks account of *@LaurenInSpace*. Typically, eBooks-style bots tweet non sequitur mashups of another user's Twitter feed, using a technique called Markov text generation (MTG).[9] The MTG approach works in quite a simple and straightforward fashion: first, we feed the generator a large amount of text with the style or content we want it to mimic or replicate. In the case of a Twitterbot, we might provide our Markov generator with a list of every tweet that the target person has ever written. The algorithm then looks at each word and makes a record of the word that comes after it. In this sentence, the word *the* occurs twice: once followed by *word* and once followed by *occurs*. For every word the algorithm discovers, it keeps a comprehensive tally of the words it finds directly after it.

When the bot sets about generating a new tweet, a Markov generator first returns to this database of words and their tallies. It randomly picks a word to start with and looks up the following-word tallies for the word it has selected. Suppose it starts with *the* and then finds these following-word tallies in its database entry for *the*:

| | |
|---|---|
| style | 1 |
| case | 1 |
| algorithm | 2 |
| word | 2 |
| occurs | 1 |

In order to pick the word that comes next, the generator must randomly choose among all of the words that were tallied. The higher the tally of a particular word, the greater the chance it has of being selected, much like how buying more tickets in a lottery increases the chance of winning. Once the algorithm picks its next word, it adds it to its sentence and uses *that* word to look up a set of tallies for the next word after that. This process continues until the algorithm has a fresh sentence to tweet. Here are some sample sentences generated by using the Markov approach to slice, dice, and tally the text of this chapter:

"Interacting with the suffix or think about home brewing might work. Even this case, 'BONG's every hour on it."

"This is important, because it never explicitly says anything, it gets away with it —and part of the Palace of Westminster in London explodes with the words 'You Matter' written in black text on it."

You can see that while these texts are far from fluent, they do seem English-*like*, and it's English of a kind we might associate with a human chatterbox suffering from cocktail party syndrome. The nouns and verbs are each more or less in the right place, even though the sentences themselves can sound strange, hilarious, or even nonsensical. Importantly, because the text is built out of words and patterns from a single source, much of the style, vocabulary, and tone of the original leak through. A Twitterbot that uses MTG to generate new tweets can often sound like a knockoff bootleg copy of the original Twitter user, which can lead to a serendipitous and surreal collision of words and ideas. Much of the time these bots produce unreadable nonsense, and our examples (generated from the text of this chapter) were sifted from many hundreds more that were unreadable. But Twitter is by its nature a terse medium, and a single crummy tweet is casually ignored as we scroll through our feeds. Finding a gem, however, can be extremely satisfying, a perfect collision of algorithms, humanity, timing, and chance. Witnessing a gem and having the sense of participating in a special moment is ultimately what makes following eBooks-style bots so much fun.

By far the most famous example of this phenomenon was *@horse_ebooks*, the Twitterbot that first gave rise to the suffix *ebooks* as a colloquial marker of bots that remix tweets from other textual sources. *@horse_ebooks* began life as an advertising bot, seemingly designed to promote e-books about horses by tweeting links to online stores. Twitter has little love for the Twitterbots that do this and is always on the lookout for accounts that might be trying to pester and spam its users with annoying links to commercial ventures. Yet there exists a whole passel of tricks to avoid detection, with one common strategy being the use of ordinary-looking text to disguise embedded advertising links. It is a gambit that worked well for *@horse_ebooks*, which would randomly select phrases from the books it was pushing to tweet alongside its commercial links.

Because of the haphazard nature of these excerpts and the fact that many become non sequiturs when robbed of their context in a book, *@horse_ebooks*'s tweets took on a very strange sheen indeed. Some would appear as reflective, calming statements ("Suddenly, I saw the beauty and wonder of life again ... I was ALIVE!") while others were much more surreal ("Make a special sauce so your dog can enjoy the festive season" or, more simply, "How to throw a horse"). The bot quickly grew from a small spam

named *@botglestats* posts a metareply to the game, containing an image filled with statistics and information about the game, including a summary of the longest words, whether any player found those words, what percentage of words that were found, and more. The two bots are by different authors (*@botglestats* is a bot by *@mike_watson*) but together they create a niche community that enhances both bots and draws their shared followers together.

If *@botgle*'s and *@botglestats*'s interplay can hardly be considered pistols at dawn, other bots show more combative spirit in their interactions. *@redscarebot* is a Twitterbot with a very clear agenda: it searches for tweets that contain words associated with left-wing politics such as *Marx* or *socialism* and then publicly quotes those tweets along with a random choice of prebaked commentary, such as "radical beatniks" or "connect the dots." The bot seems to have a jovial intent—its name is *Robot J. McCarthy*, after all—and its avatar is the infamous American politician who initiated the paranoid witch hunts that the account appears to parody. Yet in flouting bot norms, it also flaunts its disrespect for bot etiquette in a way that rubs many bot builders the wrong way and can often mark people discussing socialist politics for targeting by very real and much less jovial right-wing Twitter users. Instead of complaining, Darius Kazemi (*@tinysubversions*) took a more interesting approach: he built a bot whose only purpose is to trick *@redscarebot* into responding and muddy the waters the bot hunts in. The bot, *@redscarepot*, is named for a play on the word *honeypot*, and its tweets employ a selection of hot-button words that call *@redscarebot* to action. It offers a good example of how bots can be used to influence and play with each other in their own ecosystems.

Writing a bot to generate a statement about an issue is not uncommon; in fact, we might consider Statement bots to be another entry in our bot taxonomy, deserving of a place next to feeds, mashups, and watchers. Sometimes these bots target a very particular topic, much like *@redscarebot*, while other times they may strive to use the power of Twitter as a platform to amplify another kind of statement-making software. This mode of amplification can be remarkably powerful. In 2014, the Wikipedia article on the MH17 Malaysian airlines disaster was edited, removing text that cited Russian terrorists as the cause of the disaster and adding in its place text that shifted blame onto the Ukrainian military. The source of this edit was a computer owned by the Russian government, a fact that first came to light when a bot spied the change and tweeted it to the public. *@RuGovEdits*, by *@AntNesterov*, tracks changes made to Wikipedia articles and matches them against computer addresses that are thought to be associated with the Kremlin, tweeting out the details of any edits that match. It is part of a family of Twitterbots—from *@ParliamentEdits* in the United Kingdom to *@CongressEdits* in the United States—that aim to record and notify people when members of a country's government try to anonymously edit one of the world's most important open information repositories. These metadata are hardly a secret because Wikipedia already stores details of every edit. But identifying the source of an edit gives it a special meaning in this case, and amplifying it in public using Twitter gives the information much greater potency.

A good Statement bot need not reflect real-world data; it can also paint a counterfactual world that encourages readers to consider an alternate worldview. *@NSA_PRISMbot* is one such example of a speculative statement-making bot. Here is a representative tweet:

***FLAG*** @Okey_Robel mentioned "IRA" on Twitter. ***FLAG***

*@NSA_PRISMbot* is named for NSA's infamous PRISM surveillance program, which covertly collects and processes data about Internet use in the United States, including information about file transfers, online chats, emails, and, yes, use of social media such as Twitter. The scale, complexity, and numbing banality of the program can make the concept of mass state surveillance difficult for many of us to process, so *@NSA_PRISMbot* strives to communicate what this might mean in a different way: it tweets fictional reports about the kinds of small, everyday actions that PRISM might monitor as a way of making people think about how the very nonfictional PRISM is operating right now. Yet while *@NSA_PRISMbot* is a clever idea, it might seem that the main thrust of the message is in the idea of the bot itself, and that following it wouldn't really be any more effective or useful than, say, simply reading the previous paragraph and thinking about it for ten seconds. Nonetheless, there is an added frisson to be had when following *@NSA_PRISMbot* and its ilk, in that Statement bots sprinkle little reminders of their core message among our regular Twitter views. As we scroll past photographs of friends and idle thoughts from our favorite celebrities, we suddenly see: "Isobel Rippin of Bashirianshire, Vermont uploaded a video called DISENFRANCHISED!!! to Instagram." In this way, the message becomes a drip-feed of reminders that everything we do and everything we read may be watched by someone else. Another Statement bot, *@NRA_Tally*, operates on a similar basis, but instead of tweeting about Internet monitoring, it posts fictional mass shooting reports to which it appends stock responses from the NRA, America's pro–gun ownership National Rifle Association. The bot will contrast the horror of "11 moviego-

ers shot dead in South Carolina with a 7.62mm AK-47" with the cold indifference of a triumphalist nonapology such as, "The NRA reports a five-fold increase in membership." *@NRA_Tally* provides an interesting clash for the bot's followers to contemplate, which is not such a poor trick for a mere bot to pull off.

## Beyond the Tweet

Bots crop up everywhere, and while this book is all about the botting that gets done on Twitter, we should look everywhere for inspiration, including outside the Twittersphere. Many other sites have APIs just like Twitter that allow for automated posting, downloading of data, or accessing important functions. One particularly popular home for bots is Reddit, a vast web community of people who share links and stories and vote on which ones should earn more prominence. Bots can do many things on this site, from posting updates to submitting links and messaging users; these actions are analogous to replying to tweets, tweeting, and directly messaging users on Twitter. Reddit bots are often used as handy assistants to burnish online discussions with incidental chunks of information. For instance, one bot scans for YouTube video links, finds the top YouTube comment for that video, and appends it as a comment on the Reddit page. Another bot scans popular Reddit threads with more than five links to YouTube and creates a YouTube playlist of all of these videos, posting it along with a summary of each to the thread. Branded "Reddit's Coolest Bot," astro-bot searches for people posting photographs of space, identifies the region of space depicted and then replies with an annotated version of the image showing major stars, planets, and clusters.[14] While Twitter discussions can quickly fade because of the platform's emphasis on brief exchanges and streams of changing information, Reddit posts have, in contrast, a good deal more permanence than Twitter updates, and this allows Reddit bots to serve a longer-term purpose beyond an initial burst of comments. Reddit also hosts a panoply of bots that are designed to interject themselves into conversations, in opposition to Twitter's general guidelines (although Reddit does have its own set of API restrictions, they mostly warn against sending too many messages). The *PleaseRespectTables* Reddit bot watches for people using the "table flip" emoticon, and replies with a similar emoticon depicting someone setting the tables down normally and glaring into the screen. This bot is, sadly, now suspended in circumstances that are best described as ironic, for the bot eagerly replied too many times during a Reddit discussion that celebrated good bots.

For most websites, *bot* has been a dirty word for a very long time. Most bots have not been designed to create new meanings or new artifacts, and neither have they been gifted with a mission to help or to amuse. Many simply send unsolicited advertisements to people, while others artificially inflate the follower counts of deceptive users, allowing reprobates to sell their bot followers to others for a few fractions of a cent each. One bot written for the blogging service Tumblr allows people to automate the process of fishing for the "follow backs" that arise when users reciprocally follow those who follow them. The bot spaces out its follows through the day to avoid detection and may even unfollow people after they have followed it to improve the ratio of followers to users followed. While many of these bots automate user activities that are perfectly legal, it is hardly unsurprising that social platforms have cracked down on exploitative behaviors. Everyone has their own expectations for bot behavior, including the bot makers themselves, and Twitter is no different.

### How Not to Bot

Our whirlwind tour around the hot zones of the Twitterbot world has attempted to group bots by their behavior, the ideas behind them, or the way people enjoy them. There are a great many bots and a great many ideas for bots, and some species of bot have undoubtedly slipped through our butterfly net. We will, however, cover more bots and more bot builders in the rest of this book, and the joy of Twitterbots is that they are always evolving and showcasing new ideas. Twitterbot builders are an inventive lot, and there is always new ground to be explored, raising new questions that someone will be curious to answer. The tricky thing about breaking or exploring new ground, particularly when it concerns technology and humans mixing together in a vast public forum like Twitter, is that there are often substantial ethical issues to think about too. While bot creation might not be quite as terrifying as the stereotypical mad scientist playing God in a monster movie, letting autonomous software loose in society can have serious implications. The day-to-day world of Twitter is a tissue of fragile social situations, of people whose emotions are easily manipulated, and our Twitterbots are not always (or easily) created with a built-in sense of etiquette, good taste, or common sense. Where should the line be drawn for Twitterbot behavior? Bot builder Darius Kazemi, whom we've already met in this chapter as the creator of *@twoheadlines* and *@museumbot*, has set out some guidelines. for acting ethically as a bot author.[15] Each is worth considering in turn.

Darius Kazemi's first guideline is **don't @mention people who haven't opted in.** This is a rule that *@redscarebot* breaks every time it pesters someone for mentioning Marxism and the like. Unsolicited mentions can be annoying, since it generates notifications, and even Twitter agrees that this is bad behavior. Many bots are banned for directly messaging users who do not already follow the bot. But unsolicited mentions can do more harm than just making your phone buzz at odd times, especially if the mentions are public, as is the case with *@redscarebot*. If your bot is drawing attention to specific Twitter users, it can make those users a target for very real human harassment.

His second guideline is related to the first: **don't follow Twitter users who haven't opted in.** Though human users typically welcome human followers, the automated following of someone who did not ask for it can feel as invasive as getting pinged with unsolicited messages, and Twitter may flag this kind of behavior as poor form. Advertising bots often seek out users who tweet salient keywords so as to follow them en masse (you may have encountered this phenomenon for yourself if you have ever mentioned marketing, iPhones, or other advertising buzzwords in your tweets). So these first two guidelines are really about making sure that your bot stays within its fenced-off enclosure. We have already seen some big exceptions to these rules, though, and not all of them are as questionable as *@redscarebot*. Even though *@yourevalued* searches for users and replies to them without asking for their permission, it is hard to consider the bot a nuisance. After all, *@yourevalued* is replying to people who are arguably crying for help. Even with these simple guidelines, we can see that there is no one-size-fits-all policy for Twitterbot ethics.

Kazemi offers two other guidelines for those of us who build Twitterbots: **don't use a preexisting hashtag** and **don't go over your rate limits.** Hashtags create virtual discussion spaces where users congregate to discuss a topic, making these spaces a great place for a cynical advertiser to erect a billboard. Nefarious advertising bots thus post links with popular click bait hashtags to lure people into clicking on them, but this guideline is about more than not acting like an ad bot: it is about respecting other people's conversations and staying out of them if not invited to participate. If your bot wants to see other people, it should be interesting and fun enough to attract others to it. It should not wander over to random users like an attention-seeking toddler and foist itself into the lives and conversations of others.

Kazemi's last guideline about rate limits is another important issue for bot builders to bear in mind. Each time a Twitterbot does anything, from following a user to posting a tweet, Twitter makes a note of it. If Twitterbots do too many things in too short a time period, Twitter will slow them down, temporarily suspend them, or even permanently ban them. *Rate limiting* a Twitterbot means ensuring that the bot is sufficiently self-regulating to monitor how busy it is becoming, so that it can automatically slow itself down before Twitter takes punitive action. While this guideline is important for avoiding a shutdown, it also allows bots and their builders to show respect for the environment in which they operate. If an indifferent Twitterbot is blithely posting three hundred messages a minute, each one abusing a popular hashtag or mentioning a random user, it is quickly going to become a pest. Even if Twitter allowed such things to happen, Kazemi argues that this behavior should still be avoided. Twitterbots should be a welcome addition to a community and always aim to be on their best behavior.

Twitter has its own ideas for what makes a good bot. Some of the rules are very similar to the guidelines provided by Darius Kazemi, because Twitter is sensitive to online behaviors that look like advertising or spamming, such as sending unsolicited links to other users or repeatedly sending the same message to many different users with perhaps many different hashtags. Of course, some of its motivations differ greatly from the concerns of Twitterbot makers, and they can result in some rather peculiar decisions in the name of keeping Twitter clean. Consider the tale of two very real bots whose authors noticed the same pattern of human behavior on Twitter but responded in very different ways. The pattern in question is the rather unwise trend of users posting photographs of their new credit or debit cards on Twitter. While this might seem preposterous to some, it is not an uncommon activity: naive users often send the photos to Twitter accounts run by banks, while exuberant users may simply be showing off their brand new credit card or personalized card design. *@CancelThatCard* is a bot that automatically detects unintentionally revealing photographs using an algorithm that can identify credit cards and numbers in images. It then replies to the user with a message alerting them that their card has been seen online, with the added suggestion that they should cancel it. It even provides them with a link to a website with more information. *@NeedADebitCard* is another bot that, like *@CancelThatCard*, detects images of credit cards online. However, it makes those images even more public by retweeting them to its more than seventeen thousand followers. The account has been featured on Forbes, the Huffington Post, and tech security company Kaspersky's blog. Though currently suspended, many of the bot's retweets have prompted replies from Twitter users claiming to have ordered products online using the names and numbers revealed in the photographs.

Both bots are impudent—or educated *and* insolent if you will—though one is very clearly more malicious than the other. Nonetheless, you might be surprised by how Twitter chose to direct its ire. *@CancelThatCard* tries to quietly warn a card owner without drawing additional eyeballs to a potentially costly faux pas, while *@NeedADebitCard* seems to revel in tough love. It teaches through harsh punishment and exacerbates a rookie's mistakes by advertising them so widely. Ironically, it was *@CancelThatCard* that was first suspended by Twitter because its frequent dispatch of links to strangers, its unsolicited mentions of others, and its repeated postings of the same warnings all conspire to make its outputs read like spam. By contrast, and to a simple-minded algorithmic censor at least, the bot *@NeedADebitCard* operates with what seems like good etiquette. It retweets other users, thereby engaging in the social media world; it never follows anyone; and it never pesters other users directly. So when the detection of poor etiquette is automated, some rather strange judgments are sure to follow. Fortunately, at the time of writing, it is *@NeedADebitCard* that is suspended by Twitter, while *@CancelThatCard* continues to warn exuberant cardholders of their naïveté.

The fact that Twitter was at first unable or unwilling to act differently in the case of the two card bots shows how important it is that Twitterbot authors develop their own code of conduct; they need to think about how their bots will act, ask what kinds of rules they want to set for themselves, and decide when it is acceptable to break them. But as we have already seen in this chapter, the Twitterbot community is vast and full of diverse and interesting people. People use Twitter technology to analyze and disseminate data; they use it as a political tool; they use it as a playground for software; they use it as a canvas for art; and they use it for many more purposes and combinations thereof. This greatly complicates the question of ethics, because in the real world, different people play by different rules. Comedians are allowed to insult members of the audience, but it is much less acceptable for lecturers to insult their students.

Leonard Richardson, another Twitterbot author, explored this issue in a memorable essay, "Bots Should Punch Up."[16] He compares a Twitterbot to a ventriloquist's dummy. Although society might let the dummy say things that the ventriloquist would never be allowed to say directly, the ventriloquist ultimately takes responsibility, and so there are always lines that cannot be crossed, even by a wooden doll. As Richardson puts it, "There is a general rule for comedy and art: *always punch up, never punch down*. We let comedians and artists and miscellaneous jesters do outrageous things *as long as they obey this rule*." By "punching up" or "punching down," Richardson is referring to those who suffer at the expense of a work, whether it be comedy, art, or social commentary. Sometime the subject is obvious, as in an off-color joke about a religious group. Other times it is less predictable. *@NeedADebitCard* targets people who have made a newbie mistake by posting images of their card and its numbers online and encourages us to laugh at the mistake, or even to take advantage of it. Richardson believes this is a good example of punching down: "Is there a joke there? Sure. Is it ethical to tell that joke? Not when you can make exactly the same point without punching down."

Richardson, like many other Twitterbot builders, isn't averse to the idea that a Twitterbot can intentionally offend or provoke, and we can imagine many reasons why we might want to do this. Bots with a statement to make, like *@NSA_PRISMbot*, may well raise eyebrows or make people uncomfortable, but that is their makers' intention. Problems emerge when authors either do not consider the people their bot is affecting, such as *@redscarebot* and its focus on the left-wing politics of others, or when bots are given too much autonomy and accidentally exceed what their author intended for them. The transgression of boundaries is not uncommon, as the most creative bots are designed to do precisely this, and a bot's ability to surprise its own creator should be taken as a sign that the bot is interesting and noteworthy. This drive to build bots that can surprise us, however, is also a drive to make them unpredictable, and this can naturally yield problems in some circumstances.

Whenever we write a computer program, we naturally experience a desire to seek out useful patterns and abstractions. We have already seen ample evidence of this in the Twitterbots surveyed in this chapter. Thus, the inherent patterns of wikiHow pages allow *@wikisext* to manipulate boring English sentences—the linguistic equivalent of putting up shelves—into some delightfully euphemistic innuendo, and the reliable structure of a news headline allows *@twoheadlines* to play with cultural figureheads like so many Barbie or G.I. Joe dolls. Programming is a process that is replete with abstractions and patterns, because they yield programs that are more concise and more efficient. Yet when we try to apply the same kind of thinking to the real world, it can cause problems we may not foresee as programmers when thinking about the cold, rational world of data. The real world is not made from shiny bricks of LEGO. It has a great many gaps, bumps, and holes that are hard to imagine when we are thinking about an idea in the abstract. These pitfalls may become truly obvious only when an idea is let loose on thousands or millions of people.

## The Story So Far

This chapter has set out to provide an overview of the world of Twitterbots as it stands today. This world is a complicated mix of ideas, people, and creative potentials. Even in the time between the final editing of this chapter of the book you have in your hands today, many hundreds of Twitterbots will have come to life, covering new ground and breaking old preconceptions about what can be done with software or with the medium. At the same time, the communities of bot builders will also be evolving and updating their opinions on where the medium is headed and what its standards should be. This is both the difficulty and the beauty of writing about technology.

Twitterbotics is an inchoate technology that is still in the early stages of its development, and if you will pardon the pun, their current stage can be likened to that of another *developing* technology two centuries ago. Photography is now a staple medium of the digital age, and social media like Twitter are full to the brim with indelible visual records of the things we do, the places we go, and the people we meet. For affluent Western technology users, photography is as natural a form of communication as writing a text (and even more so than writing a letter), with apps such as Snapchat and Instagram encouraging us all to communicate primarily through this visual medium. Depending on where you are reading this chapter, there is a very good chance that you are within five feet of a camera. If you are reading a digital copy of this book, you might well be staring into one right now.

When photography was developed in nineteenth-century France, it was a very different kind of technology to that which now lets us take a snap of our dog and send it halfway across the world in less than a second. Early photography was a complicated, messy process that imposed few accepted standards other than a need for a great deal of money and time. Practitioners were forced to adopt the role of part-time chemists, experimenting with their own ways of developing photosensitive film stock. Each approach required a different combination of chemicals that were expensive and even dangerous. But as photography grew in popularity and photo subjects became photo takers, standardized approaches to taking and developing pictures emerged. The technology would soon find its way into the hands of nonspecialists such as journalists, artists, and scientists.

Once new users gain access to a developing technology and grow in familiarity with it, two interesting milestones are reached. The first is that they are soon encouraged to graduate to newer and more complex systems that build on this acquired knowledge. So today we do much more than merely take photographs: we also edit and modify them in situ to improve the way they look. Even a simple camera app on our smartphones can readjust the lighting, balance the colors, and transform the substance of our images with fancy filters. The second milestone occurs soon after we master a new technology, when we want to subvert it too. So artists don't simply use photography to replicate the world as it is; some find ways to create abstract images by manipulating light and shadow, while others use photography to freeze moments in time so they can better depict events and bend them to their own aesthetics and style. Each of these developments—the evolution of a new technology, its growth, its elaboration, and its eventual subversion—flows from having greater access to the technology, but they also go hand-in-hand with a deeper understanding of the original concept.

Generative software is currently still at that mid-nineteenth-century stage, where its practitioners mostly need to be part-time technologists—part chemist and part alchemist—to make sense of how it all works. Many bot aficionados work with their own custom-built tools, and though they may not present the physical dangers of volatile chemicals, they can certainly explode metaphorically if mixed without due care. We are entering a world where nothing is truly set in stone, and we still have no idea what generative software can do for the world or who might want to use it or what they might want to do with it. In some ways this is terrifying, and it can feel as if we are fumbling in the dark and unsure of where to go next. But it is also exciting, energizing, and a source of great optimism and joy, because every day, we can each go out and think of new ways to make systems that make meanings. We can build tools to help even more users to get involved, even if we may never be entirely sure of what we'll be doing in six months' time.

The goal of this book is to show you just one possible future for Twitterbots, just one axis along which we can develop our bots and extend their ideas and technology into something brilliant and exciting. We are going to show you how this future fits snugly alongside the many other ways that Twitterbots are being developed by other builders and how all of these strands are working together to push this medium along. We hope that in doing this, we will convince you that this world of Twitterbots is something special, something different from a silly distraction on social media, that it is in fact a blueprint for how technology and future society can integrate with one another on a larger scale.

## Trace Elements

The community of bot-builders opens its arms to all comers, regardless of programming proficiency. Even if you have never written a line of code in your life—nor have any intention of ever writing one in the future—the community provides easy-to-use tools that allow you to build and launch your own Twitterbots with a minimum of fuss. In the next chapter we will look at two of these tools, as provided by two of the community's leading lights – that reduce the task of building and deploying a text-generating Twitterbot to the specification of a grammar for the bot's textual outputs. We provide a store-cupboard of such grammars in the GitHub for this book, in a repository named *TraceElements*. In the chapters to follow you will find a section named *Trace Elements* that introduces the grammars we have placed online for expressing many of the ideas we are soon going to explore. When it comes to building bots quickly and simply, there really is no time like the present.

# 3   Make Something That Makes Something

## Animal Spirits

Twitterbots have the potential to do wild and wonderful things, from cracking jokes to making us think, but when all you can see is their outputs, it can be hard to visualize what's going on under the hood. Later in this book, we begin to build our own Twitterbots from the ground up, gluing each piece together and applying a nice coat of paint to make them shine. Before we begin, though, it is useful to think for a moment about the theory of Twitterbots and generative machinery more generally. We come across generators every day, from machines that assign school places and hospital appointments to the algorithms that drop candy from the sky in the games on our phones. In this chapter, we think about what separates one generator from another and how to think about generative software in terms of the whole of its outputs, and not just one or two examples.

To do this, we are going to make some of our own generators so we can tweak and adjust them to think about generation in different ways. To keep it simple, we are going to make our generators by hand. It doesn't take much to get going—you can just use your memory and this book—but if you have a trusty pen and paper by your side or a note-taking app on your phone, and perhaps some dice for flavor, that will certainly make things easier. We'll start by using a common generative technique that developers all over the world still use, and all it needs is a few lists of words and some random numbers.

First, we need to pick our subject matter, the domain from which we aim to pluck new artifacts. It is always useful when starting out to think of something relatively simple that has a degree of inherent structure but allows for minute variation in its details. So for our first generator, we are going to generate fictional names for fictional English pubs. In England the pubs are named after a great many things—kings and queens, historical events, guilds, aristocrats, eccentric animals, naughty farmers—but they often instantiate very simple structures that are easy to break down.[1] (You may remember a pub that was aptly named "The Slaughtered Lamb" in the movie *American Werewolf in Paris*.) Let's start by making pub names that follow the "The *<noun>* and *<noun>*" pattern, such as "The Cat and Fiddle or "The Fox and Hounds." We begin by writing down a list of ten animals. If you don't have a pen handy or prefer using a premade list, here is one we made earlier:

1. Walrus
2. Antelope
3. Labradoodle
4. Centaur
5. Pussycat
6. Lion
7. Owl
8. Beetle
9. Duck
10. Cow

We call this list a *corpus*, which is a body of text or a database of information that we can use in our generator. The list is special because we can pick any item from the list and will know something about it before we have even looked at what we picked out. We know it will be an English noun, we know it will describe an animal, and we know it will be singular, not plural. We can thus use this list to fill out our template, "The *<noun>* and *<noun>*," by randomly choosing any two entries from the list. The easiest way to do this is to flip this book open to a random page and choose the least significant digit in the page number (so if you turn to page 143, use the number 3). You can also roll two six-sided dice if you have them, summing the results and wrapping around to 1 or 2 if you get 11 or 12. Do this process twice, putting the resulting animal words into our template, and you might end up with "The Lion and Labradoodle" or perhaps "The Owl and Centaur." After a while you might want to throw some nonanimals into the list (e.g., king, queen, bishop, lord, duke) or increase the size of the list (in which case you may have to change the way you generate random numbers, but you can always generate two numbers and add them together).

In any case, you have just generated your first artifact using the list. What were the ingredients in this simple generator? Well, we simply needed a list of things to draw from, our corpus—in this case, a list of animals. Remember that we knew certain things about all the items on this list, which meant we could choose any item without knowing exactly what they were. Like dipping into a bag of mixed candy, we may not know what kind of candy we will pick out next, but we do know it will be sweet and edible, not sharp and deadly or intangible and abstract. For this particular generator, we also needed a template to fill. This template codifies everything we know about the structure of the artifacts we aim to create. Our template in this instance is very simple, but we will introduce others that will make the idea more explicit. Finally, we needed a procedure to tell us how to combine our corpus with our template. In this case, we generated some random numbers and used those numbers to choose words from our list.

Let's expand our pub name generator with some new ideas. First, we compile a new list of ten adjectives. As before, if you don't have a good way to note them down or would just prefer to use our examples, you can use our list below:

1. Happy
2. Dancing
3. Laughing
4. Lounging
5. Lucky
6. Ugly
7. Tipsy
8. Skipping
9. Singing
10. Lovable

Let's also experiment with a new template, "The *<adjective>* *<animal>*." Instead of writing *<noun>*, we've now used a more descriptive label to mark the difference between the two corpora that we've constructed. The first list is now *<animal>* and the second list is *<adjective>*. So our first template would now be rewritten as "The *<animal>* and *<animal>*." AI researchers refer to a collection of typed templates such as this as a *semantic grammar* because each template captures both syntactic and semantic constraints for generating a new linguistic artifact.[2] Using the same random number method as before, we can now choose one word from the adjective list and one word from the animal list to instantiate our simple semantic grammar and generate new pub names such as "The Dancing Antelope" and "The Tipsy Beetle."

You may have noticed while generating new artifacts that some are more appropriate as pub names than others. For example, "The Owl and Pussycat" evokes Edward Lear's poem about the two titular animals who elope in a boat and marry, defying society's standards for interspecies relationships. "The Ugly Duck" might also remind us of the children's story about an ugly duckling that transforms into a beautiful swan, reminding children that getting the last laugh is more important than being happy with who you are. A name need not reference popular culture to stand out; "The Lion and Pussycat," for example, also has a poetic quality to it, since the two animals have interesting shared qualities as well as obvious differences. By contrast, "The Walrus and Antelope" has a less obvious resonance, although this is a subjective observation and some readers might value this result over what we consider to be interesting.

Whichever you might personally favor, some results will clearly resonate with some readers more than others. But our generator is not very well positioned to understand this because it does not know that pussycats and lions are similar, just as it knows nothing of Edward Lear's poetry. In fact it does not really *know* anything, relying instead on the built-in structure of its word lists and templates. Thus, it relies on the fact that the *<animal>* list contains only animals, that the *<adjective>* list contains only adjectives, and that anywhere it is appropriate to put the name of an animal or adjective, it can choose anything from the right list.

The fact that our generators can create surprises can be both a strength and a weakness. A result like "The Owl and Pussycat" is more than just two random animals; it is also a potent cultural reference that imparts an additional flavor to the whole, making it more than the sum of its parts. Now suppose we were to show others the output of our generator but not reveal exactly how it works. They might marvel at the complexity of the generated names, asking themselves: Does our generator know how to allude to famous literary figures, or did they just write down a big list of fables and cultural ideas? They can only guess at exactly how our generator works, and in these cases, readers are often generous with their guesses. If they are presented with an artifact that carries extra layers of meaning, they might think our generator is much more complicated and clever than it really is. This unearned charity means they may really be quite impressed with our generator and even give it the benefit of the doubt whenever it slips up.

Of course, this bonus has drawbacks of its own. If our generator hits some truly high notes, followers might later feel let down if they encounter periods of noise or if they intuit how the simple system actually works.

This is a common syndrome that afflicts many AI systems, for the promise of an idea often exceeds its technical reality. Indeed, this exciting promise can curdle into disappointment once a system's details are exposed to scrutiny.

A more complex problem comes from the fact that we cannot anticipate the unintended meanings that can slip into a bot's outputs, meaning just about anything can happen. Just as our generator does not know when something is a poetic reference, it also does not know if something is inappropriate or offensive, or worse. In February 2015 police arrived at the house of Twitter user Jeffry van der Goot to discuss a death threat made on Twitter—not by Jeffry himself but by a Twitterbot based on his tweets.[3] By using Jeffry's own tweets as a corpus of word fodder to mash up into new sentences, his Twitterbot would write tweets in the same style. But just like our pub generator, Jeffry's Twitterbot doppelganger did not and could not understand the ramifications of the texts it was producing. When one bot tweet cited an event in Amsterdam among words that sounded vaguely threatening, official searches inevitably flagged the result and the police responded with haste. Jeffry was asked to decommission, or retire, the bot, which was a "death penalty" decision for the bot as at least one Twitter user saw it.

## Space Is Big—Really Big

Suppose we want to print out our pub name generator so as to give it to some friends, but we are a tad worried, or perhaps just a little curious, about what it might generate when we are not around to explain it. What we need to do is to think about our generator—or any generative system— as a whole system rather than simply considering any one output. A single output from a generator is sometimes called an *artifact*, such as, in our case, the name of the fictional pub: "The Tipsy Walrus." If we want to talk about a set of artifacts instead, we use a different term: a *space*. A space is a mathematical term for describing a set of things that share a common property or a common use or a common genesis. But we mostly use the term *space* rather than *set* because we want to emphasize that a space is something we can navigate and something we are eager to explore.

One useful space we can talk about when discussing a generative system is its *possibility space*. This is a set containing every single possible artifact a generator can produce, along with the probability of production for each one. The sum of all of these probabilities must be 1, because they represent every possible output that can emerge whenever we use our generator to make something. Possibility spaces are usually quite large. To see just how large a possibility space can be, let's calculate the size of the space of the pub name generator that uses the simple template, "The <animal> and <animal>." There are ten animals to choose from each time we want to insert an animal. Suppose we choose *Beetle* for the first animal. How many pubs are there for the partial template "The Beetle and <animal>"? That one's easy: there are ten—one for each animal that can fill the second template slot (The Beetle and Walrus, The Beetle and Antelope, and so on). The same is true if we choose *Duck* or *Labradoodle* for the first slot. So there are ten animals for the first slot, and each one combines with ten possibilities for the second slot: that makes for 10 × 10 = 100 pub names in total for this template.

Possibility spaces can grow quite rapidly as we increase the number of options to choose from. If we add an adjective to the first animal, allowing the generation of names such as "The Happy Duck and the Antelope," that increases the number of possible outputs by a factor of ten, yielding 10 × 10 × 10 = 1,000 names in total. If both animals take adjectives, there are 10,000 names in the possibility space. Note how a few small lists can conspire, via a simple template, to yield 10,000 possibilities. Large possibility spaces can be great for generators, because they facilitate the generation of a vast number of potential outputs for users to discover. If we added 100 animal names to our <animal> list and another 100 adjectives to our <adjective> list, the bot's possibility space would grow much larger, to the point where someone using our generator might never see the same output twice (assuming they only used it for a little while).

A simple case of everyday possibility spaces can be found in a deck of cards. If you take a regular fifty-two-card deck and shuffle it, it is highly likely that no one has ever produced that exact ordering of cards ever before in human history. In this case, the corpus is made up of fifty-two distinct cards from the deck, and our template is fifty-two slots, each of which asks for a different card. The mathematical calculations for this possibility space are slightly different from our animal example, because once a card has been selected, it cannot be selected again. To calculate this we use a mathematical formula for *combination*. There are 52 possibilities for the first card in the deck, then 51 possibilities for the second card (since one has been removed), then 50 for the third, and so on. Mathematicians write this as 52! or 52 *factorial*. It works out to the same value as 52 × 51 × 50 × ..., × 3 × 2 × 1.

This number is enormous, having sixty-eight digits when multiplied out, so even writing it down feels like an effort, much less imagining how big it really is. If you shuffled cards every day of your life, from the moment you were born until the day that you died, you would scarcely make a dent in this number. There are more unique decks of cards than there are stars in the universe. While that may seem like a huge space compared to our pub name generator, the mathematics of generation gets so big so fast that it's never quite as far away as you think.

### Generating with Forked Tongues

In 1941 the author Jorge Luis Borges published a collection of short stories in Spanish with a title that translates loosely as *The Garden of Forking Paths*. This collection includes the story we met in the first chapter, "The Library of Babel," in which Borges imagined a seemingly endless library of adjacent hexagonal rooms that stretch in all directions. The rooms are filled with row upon row of books, stacked up in cases, but unlike a regular library, these books are not arranged in neat alphabetical order. There is no Gardening section. You cannot find the section with all the Edward Lear poetry even if you are willing to dedicate your life to the search. Instead, each book contains a random assortment of letters, spaces, and punctuation marks, stacked shelf upon shelf, room upon room, out into the far distance. Borges's Library of Babel explores the implications of such a library if it ever really existed—and it *is* physically possible in principle—and tells extraordinary tales of the librarians driven mad by it.

We have already discussed how large the possibility space of a generator can grow, so you might suspect that the Library of Babel is *big*. But how big? Even if each book in the library were limited to just a few hundred characters apiece—this paragraph has under 400 characters—you would still end up with a possibility space so large that the number itself has more than 400 digits.

But there are many more interesting things about the Library of Babel than just its size. Although many of the books inside—indeed, most of them—are unreadable gibberish, the way the library is defined means there exist many perfectly readable books, lost among the bookshelves of nonsense. Somewhere in the library there is a perfect copy of *Moby-Dick*. Additionally, hidden away on a shelf is a copy of *Moby-Dick* that opens with the line "Call me Big Bird" instead of "Call me Ishmael." In fact, there is a copy of any book that we can conceive of that follows the rules of the library, containing only letters, spaces, periods, and commas. So somewhere in the library there is a book describing the location of every other book in the library, a perfect index catalog. (Actually, the perfect catalog would be too large to be yet another book in the library, but we can imagine a perfect catalog of catalogs that collectively describes the library.) Of course, huge numbers of other books look identical to a perfect catalog but contain subtle errors. You could spend your whole life searching for the catalog (as some of the characters in Borges's story do) and not find a book with more than a couple of words of readable English.

The Library of Babel has incredible gems hidden inside it. It even has a copy of this very book inside of it, though naturally it also has a long essay criticizing this book's authors, written by Charlemagne and Florence Nightingale. Yet these exciting finds are offset by the astronomically small odds of ever actually finding them. This is a task that often plagues the designers of generators, particularly when trying to make a generator bigger and better than ever before: How do you make your space larger and more varied without losing track of where the best bits are? When we started this chapter, we described a modest generator of pub names using just ten animal names. We knew everything about that list and how to use it, so we felt pretty confident it would generate good-quality outputs. But ten animals does not a big space make, and it can take a long time to come up with new ideas. If it seems an attractive option to use a more comprehensive list of fillers that someone else has written for us, let's try that option now.

We (and you) can use this book as our new corpus. Instead of picking two animal names from our list, close your eyes, open this book to a random page, and stab your finger down at a random word. Read ahead until you find a noun. Remember that word and then repeat the process again. Use those two words to fill our simple "The *<noun>* and *<noun>*" template, so you might, for example, generate "The List and Generator" or "The Animal and Dent." Unlike our original generator and its modest list of animals, this generator inhabits a massive space because there are thousands of different nouns in this book, many of which you might never think to use in a pub name. This is good news, because it means the generator can surprise its users more often, and it may even surprise us too, the developers who designed it. When writing this section, we expected bad names to proliferate, but "The List and Generator" was pleasantly surprising. Of course, you might notice that the generator produces many duds too. "The Animal and Dent" is not a great name. It's not even bad in an interesting way, just plain old boring, not to say incomprehensible.

Many bots rely on corpora that have not been handcrafted by their designers. If these corpora are not taken wholesale from somewhere else, they may have been automatically harvested for a particular purpose.

*@metaphorminute*, for example, generates metaphor-shaped statements by combining two random nouns and two random adjectives in a familiar linguistic template. "A gusset is a propaganda: enfeebling and off-guard" is one example of the metaphors it generates at a rate of thirty every hour (recall that it cannot quite reach the rate of one per minute because Twitter limits how often bots can use its API). These words are pulled from a large dictionary provided by *Wordnik.com* rather than a corpus designed by the bot's creator, Darius Kazemi. This combination of scale and simplicity means that *@metaphorminute* tweets every two minutes without ever worrying about repeating itself, but it also means that it is extremely unlikely that the bot will ever hit figurative gold. Like Borges's Library of Babel, we can run through *@metaphorminute*'s tweet history for months and months, getting lost in the labyrinth of its tweets, and never find anything that makes complete sense.

But is this really a problem for *@metaphorminute*? It depends on what you are looking for. If we desperately needed to name fifteen thousand pubs for a brewing behemoth and each name had to be a work of genius, then expanding our word list to include every noun in the English language would probably not help very much. But *@metaphorminute* does not aim to generate an endless stream of metaphors that are so beautiful they bring tears to our eyes. Instead, it represents a different kind of aesthetic that finds favor with a great many Twitterbot enthusiasts. Much like the tragic readers who inhabit Borges's library, running through its rooms and corridors to tear books off shelves in search of fleeting moments of surprise or elation whenever they find texts that are both readable and meaningful, the followers of *@metaphorminute* wish only to be occasionally presented with bot outputs that are mind boggling, inspiring, thought provoking, or just funny. The vastness of the bot's possibility space and its propensity for odd metaphors are intrinsic parts of the bot's character. This vastness makes the good finds more valuable and contributes to an overall feeling of exploration and discovery when we see what the bot does on a daily basis.

Just like Borges's library, each small discovery of something out of the ordinary delivers a minute thrill. For every thousand books you open, you might find just one with a readable English word. Once in every million books, you might find a readable sentence or two. If you ever found even a few hundred readable words in sequence, the experience might be so incredible and so awe inspiring that it could make the whole lifetime of imprisonment in the library seem suddenly worthwhile. Every metaphorical output by *@metaphorminute* that has even an ounce of sense to it feels special not because it is a good metaphor but because of the unlikelihood that you were there to see it when it was generated.

This again relates in some ways to the philosophical concept of the sublime—the sense of facing the vast indifference of nature and understanding our own insignificance in relation to it. The Victorians were particularly interested in the sublime's presence in nature, in how small and pointless they felt when viewing huge, untamed landscapes such as the French Alps. They experienced a feeling of horror as they came to terms with scales that they could barely comprehend, let alone measure, just as we might feel a chill as we realize the odds of ever finding a readable book in the almost infinite, dusty halls of Borges's library. Like those Victorian travelers who came back filled with existential dread, we might also feel a bit queasy contemplating being lost in such a library. Fortunately for us, we only have to worry about our generators coming up with bad names for drinking establishments. And even then, no one is forcing us to drink in those places.

### What Are the Odds?

Let's suppose that we've already opened up The Owl and Centaur in a nice forest village somewhere, amid a good-sized demographic of real ale drinkers and fans of J. R. R. Tolkien or Ursula Le Guin. With a name like that, our pub might well attract another kind of patron: players of *Dungeons & Dragons* (D&D), a pen-and-paper game that weaves interactive tales about magical adventures in fantasy worlds. For decades, D&D players have been doing precisely what we have been doing in this chapter: compiling lists of words and names and choosing among them randomly to inventively describe the parts and denizens of the worlds their games take place in. For example, to decide which languages a character speaks, a player might compile a list of languages (much as we did for animals) and roll some dice. Each roll picks out a language from the list, to randomly decide a character's proficiency in foreign tongues.

So far, our random selection processes have been just that: quite random. Flipping a book open to an arbitrary page or stabbing your finger at an arbitrary word are random processes that are almost impossible to subvert. Rolling a die is also a good source of random numbers, and because dice are small and reliable, they are popular with the folks who need a good source of randomness, like our pub-visiting, centaur-loving D&D patrons. Not all randomness is created equal, however. Suppose we come up with a wonderful new business plan for a cocktail night in which our customers

roll dice to select ingredients for their drinks. We can number our ingredients from 1 to 6 and let them roll a six-sided die (a D6) to decide what
goes into their cocktail. Assuming the die is a fair one and it is thrown
normally, each ingredient has an equal chance of being chosen.

Now suppose we expand our range of delicious generative cocktails to
include twelve different ingredients instead of six. Unfortunately we can
only afford the one die, so we simply ask each customer to roll the die
twice to select a drink. Now things are a little different; for one thing,
because the lowest a die can roll is 1, the smallest number we can get by
summing two die rolls is 2. That means that ingredient number 1 will never
be chosen. Meanwhile, we rapidly run out of ingredient number 7, because
it is now selling six times as fast as drink number 2 or drink number 12.
There are more ways to make the number 7 from two D6 rolls than any
other number in the range (rolling 1 and 6, or 2 and 5, and so forth).
Although all of these outcomes are random in the sense that we cannot
predict them in advance, they are distributed differently. We still do not
know what drink a customer is going to get, but we do know that some
ingredients and drinks are more likely than others. (Of course, if we had
a twelve-sided die, of a kind called a D12 that is often used in board games,
this issue would disappear and each ingredient would again have an equal
chance of being chosen.)

We can take this concern back to our generator of pub names too. When
we discussed ways of randomly selecting animal names, we suggested flipping to a random page and choosing the least significant digit from the
page number (so page 145 would give a 5). Now suppose that instead of
using the least significant digit, we choose the most significant digit. What
happens to our animal generator now? Well, now the tenth animal now
will never be chosen, because page numbers never start with a zero. The
remaining nine are still all possible outcomes, but there are many more
ways to choose the first few animals in the list than the rest. To understand
why, consider the distribution of page numbers in this book. There are
nine pages with single digits (pages 1 to 9), which all provide an equal
chance of choosing any digit. Then there are 90 pages with two digits,
which are also equally distributed between 1 and 9. So up to this point,
there are eleven pages that start with any of the numbers in the list. But
because this book has fewer than a thousand pages, all of the remaining
pages start with a 1, a 2, a 3, or a 4, and there are almost 100 of each of
these pages, which greatly biases our selection. In total, there are 111 pages
that begin with a 1 in this book, but perhaps only 11 that begin with a 9.
All of this means that you're much more likely to see *Labradoodle* in your

generated pub name than you are to see *Owl*, because *Labradoodle* sits higher in our list of animals.

While the quirks of these strategies are well known to those who use them a great deal, such as the probability distribution of two summed six-sided dice (if you have ever played the dice game craps, this is where the various betting odds come from), other kinks in a probability distribution are easier to forget, such as the difference between the most and least significant digits in the page numbers of a book. Not many generators use page numbers as their source of randomness, but this uneven probability distribution can also rear its head in the corpora we use just as much as in our random selection methods. If we delete the word *Labradoodle* from our animal list and add a second *Lion* in its place, then Lion is twice as likely to appear as any other animal on the list. Much like many of the ideas we have touched on in this chapter, this is an easy fact to recognize when the list has just ten animals. But when we start to look at huge corpora with hundreds of thousands of items apiece, we can easily forget.

The *@guardiandroid* bot mashes together opinion column headlines from the British newspaper the *Guardian* to invent fake headlines that sound like real ones. It uses much the same mashup technique as Jeffry van der Goot's bot mentioned already in this chapter, the one that almost got him arrested. (*@guardiandroid* has not caused anyone to be arrested at the time of writing.) That is, it chops up existing headlines and rearranges their various words and phrases according to the dictates of statistical analysis to manufacture new ones that retain much of the flavor of the originals. This approach can sometimes work surprisingly well, and humorous combinations of topical issues often emerge. One fake headline cries, "Michael Gove needs to be shaken up" (Gove was the UK education secretary at the time of tweeting, and a person that many would like to shake vigorously), while another pleads, "Forced marriage is a deeply malign cultural practice—but it's not only killer whales we must protect."

The problem is that some opinion headlines regularly use the same linguistic patterns, which means that when our bot automatically searches for column headlines, these recurring patterns will emerge multiple times from our corpus, rather like our reuse of the word *Lion* in our animal list. For example, the *Guardian* ran a regular series of columns based on the common questions people ask Google. Each headline began with a question and ended with the phrase, "You asked Google—and here's the answer." Because the phrase was repeated so often throughout the corpus, the bot created a disproportionately high number of headlines with this exact phrasing, since it is more likely to choose headlines with this pattern than with any other. Once we noticed this abnormality, it was relatively easy to fix by removing surplus references from the corpus. But noticing it in the first place is difficult. Often these strange bumps in our work are evident only when they appear in public-facing outputs.

Issues of probability distribution can creep into systems in strange and subtle ways, and this can make them especially hard to detect and address. Sometimes a specific probability distribution is used intentionally, as many games (such as Blackjack) hinge on an understanding of probability theory. As we shall see when building our own generative systems in this book, stumbling across the unusual outcomes of our Twitterbots is all part of the fun of building them, but knowing about these issues in advance can help us to know what to watch for and what common mistakes to suspect when something goes wrong.

### Bot the Difference

So far, we have gained some familiarity with piecing texts together with templates and thinking about the weirdness that can come from generating at large scales when the text is uncontrollably big and the generator can make more things than we could ever review in a lifetime. We have also looked at a powerful driving force behind simple generative systems—random number generation—and how the probabilities of different results can be affected by so many different issues.

As designers of generators (which we all now are, with our little pub name generator behind us), one common way to test what our generator does and how it is performing is simply to use it to generate some outputs. We might generate one or two pub names and look at them. Do they look okay? How bad are they? If we are feeling more thorough, we might generate a few hundred. Are we bored by them? Are there any surprises? Any repetitions? Perhaps we can go much further than a hundred. Maybe we should keep generating pub names until we fall asleep with boredom or fatigue. How many should we look at before we stop?

Recall that earlier, when calculating just how large a generative space can be, we considered what would happen if we gave our simple generator a few more choices and a more ambitious template. Our template "The *<adjective> <animal>* and *<adjective> <animal>*," when filled using just 20 adjectives and 20 names, provides 160,000 ($20 \times 20 \times 20 \times 20$) potential outputs. That sounds like quite a large number. In fact, it seems so big we might want to boast a bit to our friends: 160,000 different names for pubs from a few simple lists and some tricks! That's three times more pub names than the number of pubs operating in England right now, some of which have less-than-desirable names (and beers to match).

When we describe our pub names as "different," we allow everyone to have their own personal interpretation of what that word might mean. To some, it means that they are as different and varied as real English pub names such as The Cherry Tree, The Duke of York, The White Hart, and The Three Crowns. To others, it might mean that they have the same overall appearance, but the words are varied and changing. They might imagine our generator having a list of hundreds of animals and adjectives to pick from instead of just twenty of each. To us, knowing how the generator works, we know that what we mean is simply that no two pub names are exactly alike. Each name has something different about it, even if it is just the order of the words. The Laughing Beetle and Tipsy Walrus is, strictly speaking, different from The Tipsy Walrus and Laughing Beetle. But it's not as different as The White Hart is from The Duke of York.

The language we use to describe our generators is complicated and is often not given as much thought as it deserves. The video game *Borderlands* was released in 2009 with the claim that it contained 17.75 million generated guns, which excited many players at the thought of endless variety and discovery. In 2016, *No Man's Sky* promised 18 quintillion planets to explore. But if many of those guns and planets are, for all practical purposes, indistinguishable from one another, these numbers are neither exciting nor truly meaningful. Twitterbot author and generative expert Kate Compton (@*GalaxyKate*) colorfully names this problem "The Ten Thousand Bowls of Oatmeal" problem.[4] Yes, you can make ten thousand bowls of oatmeal for a picky customer, so that every single bowl has a uniquely specific arrangement of oats and milk, with an array of consistencies, temperatures, and densities that would flummox the choosiest Goldilocks. But all that our picky customer is actually going to see is an awful lot of gray oatmeal.

So what's going on here, and how can we tackle this problem? The main issue is that there is a real difference between the technical, mathematical notion of uniqueness and the perceptual, subjective notion that most of us understand. The bowls of oatmeal may be unique at a very low level, but what matters is what we can perceive and remember. We might want, in an ideal world, to build a Twitterbot that always produces completely distinct concepts, but this is an unrealistic goal to aim for, even for our own creative oeuvres as humans. Instead of worrying about how to always be different, it is just as productive to focus on ways to avoid becoming overly familiar. For example, we might build our pub name generator to remember the animals and adjectives it has used recently so as not to reuse them until the entire list has been worked through. That way, a more memorable word like *Labradoodle* appears again in a name only after all the other animals have been used and the system resets itself. Compton calls this approach *perceptual differentiation*. Readers may remember with time what animals are in our generator, but as long as similar outputs do not recur too close to each other, it might take longer to appreciate the limits of the system.

Another easy way to refresh a generator is to add more content to its corpora over time. We have already seen how adding just a few extra words to a list can significantly increase the number of possible outputs. Adding to a generator later in its life disrupts many of the patterns that its users will have become familiar with. This is good not only insofar as it adds interesting new outputs to the bot's repertoire, but because new patterns can also break readers' expectations of what a bot is capable of doing. Many bot builders add to their corpora over time, as in the case of a bot named *@yournextgame* by *@athenahollow* and *@failnaut*. New injections of content into a bot's corpus can significantly extend the bot's life span and enable unusual new interactions with the existing words that the bot has already worked with. For many authors, including *@yournextgame*'s creators, adding to a bot is an especially pleasant kind of maintenance task, much like tending to a secret garden or maintaining a beloved car. It offers us a chance to engage in a cycle of creative reflection, a chance to think about why we made a bot in the first place, and a chance to think about how we might make it better.

Differentiating your bot's oeuvre also depends on where in the bot its variety and scale actually derive from. Suppose that instead of having twenty adjectives and twenty animals in our pub name generator, we have in fact only one adjective and one animal: *Dancing* and *Cat*. But this time instead of one template, we have 160,000 pub templates, each with slots in them for just one animal, or one adjective, or both. This list is a little hard to imagine, because it would require so much effort to compile, but we might expect that this generator is a good deal more memorable than one that uses a single template and replaces the animals and adjectives. (If Borges wrote a story about zany Twitterbot builders, he might use this as a plot hook.) The structure of a pub name is much more memorable than any single noun or adjective it contains, so changing that structure creates a greater sense of variation in the output. Our lists of animals furnish additional detail to a template, allowing readers to ponder what, for example, a walrus has to do with a beetle. Yet over time, relentless repetition of the same template can wear down its novelty, leaving us overly familiar with its slots and feeling ho-hum about their changing fillers. In contrast, tens of thousands of templates would be that much more striking, for as the detail stays the same, the wider structure is ever shifting and offering up new patterns for us to consider.

Meaninglessness is not always a thoroughly bad thing. We have already seen in this chapter that even empty generation can sometimes be a virtue of sorts, inasmuch as it offers us the same joy that is felt when discovering something readable in the Library of Babel after many years of search. There is no obviously right way or wrong way to make a generator, so when we talk about making this generator meaningful, more or less, we are really talking about our own personal goals for what a generator should be capable of communicating to its followers. Yet understanding other people's expectations for our software is important, and probably more important than understanding how to code in Java and Python or how the Twitter API works. If followers do not know what to expect from your bot, they might well hope for too much, and that way brutal disappointment lies.

**Home Brew Kits**

The mechanics of template filling are easy to specify, especially when we want to fill the slots of our templates with random values from predetermined sets. To ease the development of our template-based generative systems, which always revolve around the same set of core operations—pick a template, pick a slot, pick a random value to put in a slot, repeat until all slots are filled—Kate Compton, whom we met earlier when musing about the perceptual fungibility of oatmeal, has built an especially useful tool, Tracery, for defining simple systems of templates and their possible fillers.[5] Indeed, Tracery has proven to be so useful for template-based generation that another prominent member of the bot community, George Buckenham, has built a free-to-use web service named Cheap Bots Done Quick (CBDQ) around Tracery that allows bot builders to quickly and easily turn a new idea for a template-based bot into working reality. His site, *CheapBotsDoneQuick.com*, will even host your template-based bot when it is built, sidestepping any programming concerns that might otherwise prevent your good idea from becoming reality. For CBDQ to host your bot, you will also need to provide the site with the necessary tokens of your bot's identity that any Twitter application will need before it can tweet from a given account and handle. However, Twitter makes registering any new app—such as your new bot—easy to do (we walk you through that process in the next chapter). CBDQ simplifies the registration process even further, to a simple button click that allows it to talk to Twitter on your behalf.

CBDQ helps developers rapidly create new template-based Twitterbots by setting down their patterns of generation in a simple specification language, one that is remarkably similar to the templates we have used for our pub names and phrases in this chapter. The approach is not suited to complex bots that require a programmer's skills, but Kate devised the Tracery system as a convenient way of describing a context-free generative process that is fun, simple, and elegant. As an example, let's see what our pub name generator looks like when it is written out as a set of Tracery templates or rules. At its simplest, our generator needs to put two animal names into each template to make a name, or it needs to marry a single adjective to a single animal word. So in Tracery we define the following:

```
"origin": ["The #animal# and #animal#,"
           "The #adjective# #animal#"],

"animal": ["Walrus," "Antelope," "Labradoodle," "Centaur,"
           "Pussycat," "Lion," "Owl," "Beetle," "Duck," "Cow"],

"adjective": ["Happy," "Dancing," "Laughing," "Lounging,"
              "Lucky," "Ugly," "Tipsy," "Skipping," "Singing,"
              "Lovable"]
```

Tracery is a *replacement grammar* system. A replacement grammar comprises a set of rules that each have a *head* and a *body*. In Tracery, the head is a single-word label, such as origin or animal, and the body is a comma-separated list of text phrases. Tracery begins with a start symbol (e.g., origin), which is the head of a top-level rule. It then finds the rule body that matches that head and randomly selects a possibility from its body list. So a symbol like animal can be processed by Tracery to pick out a replacement term like walrus or beetle.

In some rules, you will notice that words are bracketed by the hash symbol #. These keywords denote the points of replacement in each template. So when Tracery encounters one of these special keywords, it checks to see if it matches any of the heads of any other rules. If it does, it goes off (recursively) and processes that rule to replace the keyword with the result of processing the rule. We can see this in action if we choose the start symbol "origin."

```
"origin": ["The #animal# and #animal#,"
           "The #adjective# #animal#"],
```

Notice that each rule in our Tracery grammar, except for the very last, is followed by a comma. The commas in each list denote disjunctive choice: we can choose either of the above expansions for "origin" (to yield either a name with two animals or a name with a single adjective and a single animal). Suppose that Tracery chooses the second option, "The #adjective# #animal#." When Tracery processes this and finds the keyword "#adjective#," it then goes to the rule with the head "adjective":

```
"adjective": ["Happy," "Dancing," "Laughing," "Lounging,"
             "Lucky," "Ugly," "Tipsy," "Skipping," "Singing,"
             "Lovable"],
```

It selects a word from the body of the rule and goes back to the original pattern and replaces the keyword with this new word. This process is then repeated for the second term "#animal#" in the phrase. When Tracery finishes replacing all of the keywords, the phrase is complete and it can be returned, and so we get back "The Laughing Centaur" as our pub name. So in just three lines, we can rewrite our entire pub name generator. Moreover, the Cheap Bots Done Quick website can take this specification and turn it into a fully fledged Twitterbot. All we need to do is allow CBDQ to register our bot with Twitter so CBDQ can use its permissions to post its tweets, and we're done. The simplicity and effectiveness of CBDQ has led to many popular bots being made using this site. It's also a favorite for teachers looking to introduce students to generative software in a quick and snappy manner. Think of CBDQ as a support system for your ideas; if you can invent a new idea for a bot in which everything that goes into the bot's tweets can be expressed as a series of templates—and bots such as *@thinkpiecebot* and *@LostTesla* show that quite sophisticated tweets can be generated using a well-designed system of templates—then Tracery and Cheap Bots Done Quick may be the tools for you.

## Just Desserts

It is tempting to assume a clear distinction between the bots that we can build using Tracery/CBDQ and those that require us to write code in a conventional language such as Java or Python, but there is no reason we cannot do both. A principal reason for taking the programming route is that Tracery grammars are context-free, so that substitutions performed in one part of a rule do not inform the substitutions performed in the other parts. Context-sensitive grammars are heavy-duty formal devices that allow substitutions to be performed only on the parts of a structure that are appropriately bracketed by specific substructures, but Tracery is wise to avoid this route.[6] In many cases, a concise context-sensitive grammar can be rewritten as a much larger—yet still quite manageable—context-free grammar by expanding a single context-sensitive rule into many context-free rules that capture the very same cross-dependencies. If this seems like a tiresome task that makes the programming route a compelling alternative, there is a middle ground: we can write a small amount of code to do the conversion from context-sensitive to context-free for us. That is, we can write some code to take a large knowledge base and generate a Tracery grammar that respects the contextual dependencies between different kinds of knowledge.

Suppose we want to build a bot that generates novel recipes by substituting one element of a conventional recipe for something altogether more surprising. AI researchers in the 1980s found the culinary arts to be a fruitful space for their explorations in case-based reasoning (CBR), an approach to problem solving that emphasizes the retrieval of similar precedents from the past and their adaptation to new contexts. Janet Kolodner's JULIA[7] (named for famed chef Julia Child) and Kristian Hammond's CHEF[8] both explored the automation of culinary creativity with CBR systems that could retrieve and modify existing recipes to suit new needs and fussier palates. For simplicity we focus here on the invention of new dessert variants with an unpalatable twist: our bot, and our Tracery grammar, is going to generate disgusting variants on popular treats to vividly ground the oft-misused phrase "just desserts" in culinary reality. "Just deserts," meaning that which is justly deserved, is an Elizabethan coinage that is frequently misspelled[9] as "just desserts" on Twitter and elsewhere, in part because we so often conceive of food as both a reward and a punishment. Our approach will be a simple one and use a Tracery grammar to generate a single substitution variant for one of a range of popular desserts. Let's imagine a naive first attempt at a top-level rule, named "origin" to denote its foundational role (in Tracery we name the highest-level rule "origin" so that the system always knows where to start its generation-by-expansion process):

```
"origin": ["#dessert# made with #substitution# instead of
           #ingredient#"].
```

So a new dessert variant can be generated by retrieving an existing dessert (such as "tiramisu"), an ingredient of that dessert (such as "mascarpone cheese"), and a horrible substitution for that ingredient ("whale blubber," say) to generate the variant recipe: "Tiramisu made with whale blubber instead of mascarpone cheese." But notice that the expansions performed here are context-free. The grammar is not forced to select an #ingredient# that is actually used in tiramisu, nor is it constrained to pick a substitution that is apt for that specific ingredient. It might well have chosen "pears" for the ingredient and "spent uranium" for the substitution, revealing to the end consumer of the artifact that the system lacks any knowledge of cookery, of either the practical or wickedly figurative varieties. To make the grammar's choices context-sensitive, we need to tie the choice of ingredient to the choice of dessert and the choice of substitution to the choice of ingredient. We can begin by defining a rule that is specific to tiramisu:

```
"Tiramisu": ["Tiramisu made with #Marsala wine# instead of
            Marsala wine," "Tiramisu made with #mascarpone
            cheese# instead of mascarpone cheese," "Tiramisu
            made with #dark chocolate# instead of dark
            chocolate," "Tiramisu made with #cocoa powder#
            instead of cocoa powder," "Tiramisu made with
            #coffee powder# instead of coffee powder,"
            "Tiramisu made with #sponge fingers# instead of
            sponge fingers"],
```

This rule is the go-to rule for whenever we want to generate a horrible variation of tiramisu for a guest who has overstayed his welcome. Notice that it defines one right-hand side expansion for every key ingredient in the dessert. These expansions in turn make reference to subcomponents (called *nonterminals*) that can be expanded in turn, such as #coffee powder# and #mascarpone cheese#. The following rules provide wicked expansions for each of these nonterminals:

```
"coffee powder": ["black bile," "brown shoe polish," "rust
                  flakes," "weasel scat," "mahogany dust,"
                  "baked-in oven grease"],

"mascarpone cheese": ["plaster of Paris," "spackle," "mattress
                      foam," "liposuction fat"],
```

These rules make the substitutions of ingredients context-sensitive: mahogany dust may be chosen as a horrible substitute for coffee powder but never for egg white. We now need a top-level rule to tie the whole generative process together:

```
"dessert": ["#Almond tart#," "#Angel food cake#," "#Apple brown
            betty#," "#Apple Charlotte#," "#Apple crumble#,"
            "#Banana muffins#," . . . , . . . , "#Vinegar pie#,"
            "#Vanilla wafer cake#," "#Walnut cake#," "#White
            sugar sponge cake#," "#Yule log#," "#Zabaglione#"],
```

To generate a horrible "just dessert," the grammar-based generator uses the dessert rule above to choose a single dessert as an expansion strategy. The rule that corresponds to that dessert is then chosen and expanded. Since that rule will tie a specific ingredient *of that dessert* to a strategy for replacing that ingredient, the third and final level of rules is engaged, to

```
"other": ["Mexican," "foreign," "migrant," "Chinese,"
          "un-American," "Canadian," "German," "extremist,"
          "undocumented," "communist," "Dem-leaning," "freedom-
          hating"],

"worker": ["car-makers," "laborers," "cooks," "workers,"
           "radicals," "office workers," "journalists,"
           "reporters," "waiters," "doctors," "nurses,"
           "teachers," "engineers," "lawyers,"
           "mechanics"],

"praise_target": ["current and ex-wives," "FOX & Friends
                  buddies," "supporters," "McDonald's
                  fans," "coal miners," "hard-working
                  employees," "business partners,"
                  "investors," "diehard Republicans,"
                  "alt-right wackos," "friends on the
                  hard-right," "legal children," "human
                  children"]
}
```

Once entered into the Tracery window in CBDQ, you can test your grammar by asking the site to construct sample tweets, such as the following:

Those Covfefe-sucking media clowns have bad hair

CBDQ also provides some decent error checking for your Tracery grammar as you write it, allowing developers to craft their bots incrementally within the cozy confines of the site, much as programmers develop their code inside an interactive development environment (IDE). Even if you graduate to the complexity of the latter to build bots that go beyond the limitations of simple context-free grammars, it pays to keep one foot firmly planted on CBDQ soil. Your stand-alone bot code may do all the running when it comes to complex outputs that tie the different parts of a tweet together in a context-sensitive fashion—imagine a bot that poses analogies and metaphors that coherently relate a subject to a theme—but this code can comprise just one string in your bot's bow. In parallel, a CBDQ version of your bot may quietly tweet into the same Twitter timeline, so that the outputs of your bot come simultaneously from two different sources and two different mind-sets. Indeed, if your stand-alone code should ever fail, due to a poor Internet connection, say, you can always rely on the CBDQ/Tracery component of your bot to keep on tweeting. Thus, the Tracery grammar above is one part of the functionality of a bot we call *@TrumpScuttleBot* (ClockworkOrangeTrump).[10] The stand-alone side, written in Java, generates tweets that give the ersatz president an ability to spin off-kilter analogies about the media and other rivals, while the CBDQ side-lobs a daily tribute or insult over the same Twitter wall. We also use CBDQ to respond to any mentions of *@TrumpScuttleBot* in the tweets of others, so that the Java side can devote all of its energies to figurative musings.

Bot builders with clever ideas that fit snugly into the context-free confines of a Tracery grammar may see little need to move to a full programming language and may not be inclined to see such a move as a "graduation." Over time, such bot builders may accumulate a back catalog of bots and grammars that is extensive enough to lend a degree of playful serendipity to the development process. Those past grammars are not black boxes that can be reused as self-contained modules, but cut-and-paste allows us to easily interweave old functionalities with the new. Consider, for instance, the possible relationship between a *just desserts* grammar and a Trump grammar: our bot president might use the former when hungrily musing about the right way to devour those in his digital cross hairs. If we cut-and-paste all but the `origin` rule from the desserts grammar into our new Trump grammar, It just remains for us to add an expansion like the following to `origin`:

```
"I will make those #insult_adj# #insult_target# eat #dessert#,
believe me"
```

We have seen a variety of bots whose modus operandi is to cut up and mash up textual content from other sources, such as news headlines and the tweets of other Twitter users, but the cut-and-mash approach can apply just as readily to the DNA of our bots as to the specific outputs they generate. When we think of Tracery grammars and other means of text generation as defining flavors of text, our bots become free to serve up as many scoops as will fit into a single tweet.

### How Much Is That Doggerel in the Window?

Every generator is, in some sense, unique. It's uniqueness depends on who is making it, what they are making it with, and what they are making it about. Ask a few friends to write down their ideas for pub names, and each will approach the task from a slightly different angle. The same goes for generators of all stripes: a variety of techniques, languages, inputs, and outputs can all be used to produce generators of different sizes, complexities, and output styles. This means that there is no one right way to build a generator, but we can surely help out with some general guidelines to guide you to where you want to go.

Every generator has a starting point. Sometimes it's an idea for creating a new kind of artifact in an already crowded space of interest, such as a bot that lampoons a strident politician or invents new ideas for video games. Sometimes it is an amazing source of data that you have just stumbled on, such as a list of a thousand video game names,[11] or political donations, or common malapropisms, or a knowledge base of famous people and their foibles (we provide this last one in chapter 5). One of the best ways to start thinking about a new generator is to create some inspiring examples by hand. Creating examples manually allows you to notice patterns in the content that you might not otherwise appreciate. These patterns need not generalize across all of the examples that you produce, but you might notice two or three with something interesting in common. Thus, while most pub names do not obey the pattern "The *<something>* and *<something>*," enough do so that they stand out when you write some down or look some up on the Internet. Patterns can help you to pick apart the layers of abstraction in your content to see what changes, what stays the same, and by how much it varies. Identifying the moving parts of what you want to generate will help you to see which parts can be extracted and handed over to an automated generator.

So let's build a new generator by starting with an example of what we want to generate and thinking about how to gradually turn it into a generative system. This generator, like our generators of pub names and Trumpisms, will be another example of a grammar-based generator. Indeed, the term *grammar* can be used to label any declarative set of rules that describes how to construct artifacts from their parts, whether those artifacts are texts, images, or sounds. For our pub name generator, our rules defined a simple semantic grammar for describing how to name English pubs (by filling slots such as *<animal>* with the results of other rules) and how to choose an animal name (even if trivially this just means plucking it from a fixed basket of options). Grammars are especially suited to the generation of content that is highly structured and easily dissected into discrete recombinant parts. For our pub names, every name employs the same kinds of element in the same places, so that we can easily pull out the *<animal>* parts and the *<adjective>* parts and write lists to generalize them. Our Trump bot explores a world with very different flora and fauna, but it does so in much the same way.

Let's explore a popular love poem format, often called "Roses Are Red," whose template structure yields poems that are disposable but fun. The poems are short, constrained, and heavily reliant on rhyme. Here is a typical instance:

Roses are red
Violets are blue
Sugar is sweet
And so are you.

There are a great many versions of this poem and many jokes that use the same poetic structure. Each one varies a little from the basic version above. Some variants change words or entire phrases, while others change the meaning of the poem or play with its structure. Here are a few more examples, so that we can begin to look for the common patterns that we can feed into our generator:

Roses are red
Violets are yellow
Being with you
Makes me happy and mellow.

This poem has the same sentiment as the first, but the poet has changed the rhyme in the second line by switching in a different color, in order to end on a different note. It also removes the comparison in the third line. Here is one more:

Roses are red
Violets are blue
Onions stink
And so do you.

These variants only slightly tweak the poetic format, which suggests a Tracery grammar could be used to capture the variability in form. We can also see some obvious points of replacement where we might insert some lexical variability. We can change the colors or properties of the poetic descriptors (sweetness or redness) and the things being described (sugar or roses). We can change the sentiment of the last line or the way it is expressed. So let's start with something simple and swap out some of the words in the first two lines. *Roses* and *Violets* both play the same role in the poem: each denotes an object in the world that the poem will say something about. So to find something to replace them with, we must find what they have in common. They are both words, so we could just compile a list of words to replace them, and then choose randomly:

Yodeling are red
The are blue
Sugar is sweet
And so are you.

This poem fails to scan, though that is the least of its issues. We have also broken some English grammatical rules by being too broad in our word choice. We can be a bit more specific though. *Roses* and *Violets* are both plural nouns, so we can compile a list of plural nouns to replace them, to yield the following variant:

International trade agreements are red
Cherries are blue
Sugar is sweet
And so are you.

This approach is not all that bad, and the poem at least makes some sense. The advantage of this approach is that lists of nouns are easy to compile and just as easy to turn into the right-hand side (the expansion side) of a Tracery grammar rule. There are thousands of nouns we can use, and it is often very easy to find lists that we can simply copy and paste into our bot. Crossword- and Scrabble-solving sites often have word lists we can download, and we might even be able to get lists directly from an online dictionary, much as *@everyword* did. These poems do not always make complete sense, however, so if we wanted, we could look for more specific connections. The purpose of using *Roses* and *Violets* in the poem is that each is a flower and flowers are romantic, so we could compile a new list of flowers to replace our general nouns. This might yield this variant:

Azaleas are red
Chrysanthemums are blue
Sugar is sweet
And so are you.

Now these are much closer to the original poem, because our choice of words has semantic value to it, just as our list of animals worked better than a simple list of random nouns for our pub name generator. However, any gains we make in cohesive meaning come at the expense of surprise and variety. There are fewer words for flowers than there are nouns since the former is contained within the latter. Because we are tacitly embedding meaning into our word lists, readers will eventually adapt to the new outputs and realize that they will never see anything other than a flower in those first two lines. This is a trade-off that is fundamental to much of bot making, particularly to those bots that rely on grammars. Bot builders must decide how much consistency and high quality they are willing to trade off against surprise and variety. We could get even more specific if we so desired. "Chrysanthemums" is an awkward word that disrupts the meter of the poem, so we might compile a list of flowers whose names have just two syllables apiece. But this produces an even smaller list, leading to less variety and more repetition, even as it potentially gives us poems that sound better and look more similar to the original. So it becomes a matter of personal taste as to how far along this line a bot builder is willing to go.

We can do the same for the color words. First, we might use our original random word list, or we could compile a new list of adjectives, or a smaller list of adjectives that make sense for flowers (this list might include "beautiful" but omit "warlike" and "gullible"). We could also compile a list of colors of different syllable counts, such as a list of monosyllabic colors. Yet even this might not be enough:

Lilies are gold
Poppies are puce
Sugar is sweet
And so are you.

We have not even touched the last two lines, so now the poem has lost its rhyme. We could leave it like this and wait for rhyming words to naturally appear in the generator's output, accepting that sometimes the bot's output will rhyme and other times it will not. We could aim to extend our word list even more by including only colors with one syllable that also rhyme with "you", but that list is likely to be extremely short. We could also sacrifice some features in exchange for others, such as removing our color constraint altogether and compiling a list of general adjectives to give us a larger pool of words to work with. For instance, we might use only the adjectives that rhyme with "you," as in:

Lilies are gold
Poppies are new
Sugar is sweet
And so are you.

So we sacrifice one kind of quality for another. This sacrifice can make many forms, and we might use the opportunity to make our poems more pointed by reusing parts of earlier bots and their grammar base. Suppose we wanted to reuse parts of our Trump bot for at least some variants of our poems. Instead of using word lists to provide the nouns, such as flowers (lilies and poppies and roses and violets) and the evocative descriptors (sugar, honey), we could reuse the flattering and insulting adjectives, nouns, and behaviors that form the basis of our ersatz president. We still have the problem of rhyme: if the last word of the poem is "you," then the end of line two must also rhyme with "you." This is such a common ask that *Rhymer.com* provides a list of words that rhyme with "you."[12] The words span all syntactic categories and cannot simply be dropped into the last word position without some semantic mediation. But we could take the bulk of these words and build a Tracery grammar rule to use them meaningfully:

```
"praise_you": ["#praise_action# too," "look good in a
               tutu," "whipped up a great stew," "love a cold
               brew," "Trump steaks do they chew," "on Trump Air
               they once flew," "always give me my due," "jump
               when I say BOO," "never wear ties that are blue,"
               "hate cheese that is blue," "my praise do they
               coo," "supported my coup," "work hard in my
               crew," "my bidding they do," "Russian pay checks
               they drew," "never bleat like a ewe," "never gave
               me the flu," "number more than a few," "rarely
               sniff cheap glue," "asked me to pet their gnu,"
               "never cover me with goo," "my bank balance they
               grew," "to no ideology they hew," "never laugh at
               my hue," "hate liberals, who knew?," "always
               bark, never moo," "to Washington are new," "like
               to kneel at a pew," "in gold toilets will poo,"
               "for my inauguration did queue," "my opponents
               will rue," "are not short of a screw," "would
               lick dirt from my shoe," "from the Democrats will
               shoo," "never vote for a shrew," "run casinos
               (they're 1/16th Sioux)," "to the right they will
               skew," "a fearsome dragon they slew," "no leaks
               do they spew," "testify when I sue," "an
               electoral curveball they threw," "know to them I
               remain true," "suggest women I can woo," "worship
               me as a guru," "made me king of the zoo," "can
               count to two"]
```

We can now define a Tracery rule to create the second line of our poem:

```
"line_two": ["#praise_target# #praise_you#"]
```

This will generate lines such as "diehard Republicans never laugh at my hue." It just remains for us to define rules for the other three lines in much the same way. In fact, we might add the poetry generator as an extra dimension to our satirical Trump bot, simply by defining this additional expansion for that bot's "origin":

```
"Roses are red, violets are blue, my #praise_adj#
#praise_target# #praise_action#, and #praise_you#."
```

By tapping into existing Tracery rules, this will produce poetic nuggets such as:

> Roses are red,
> violets are blue,
> my FANTASTIC friends on the hard-right are HIGH energy,
> and made me king of this zoo.

We can now crack open the boilerplate of the first two lines by defining expansion points for different kinds of flowers, or at least alternatives to roses and violets:

```
"red_thing":   ["Roses," "Southern states," "Bible belt
               states," "Trump steaks," "Chinese-made ties,"
               "McDonald's ketchup," "Rosie O'Donnell's cheeks,"
               "Megyn Kelly's eyes," "#MAGA hats," "Trucker
               hats," "Tucker Carlson's bow ties," "fire trucks,"
               "bloody fingerprints],

"blue_thing":  ["violets," "Democrat states," "NYPD uniforms,"
               "Hillary's pantsuits," "secret bus recordings,"
               "West Coast voters," "East Coast voters," "secret
               Russian recordings," "Smelly French cheeses"]
```

As an exercise, why not add some additional colors to the poem template? If "orange" is too challenging (it *is* one of the hardest words to rhyme), then how about "gold"? Online poetry dictionaries suggest many rhymes for "gold," so all you need to do is define a rule for "**gold_thing**" and another for "**praise_gold**."

But suppose we want to change the last line of the poem too, using a list of words to replace "you" with plural nouns such as "aardvarks," "politicians," or "leaks." Now we have unfrozen two moving parts that need to match up and rhyme, but suddenly it is all becoming just a little bit too complicated to do by randomly choosing words from lists or, for that matter, choosing expansions randomly from a context-free grammar rule. This is where building a bot in a programming language can help us. For example, if we supply our bot with access to an online rhyming dictionary, it can look up the words that rhyme with our adjective on the second line to find a suitable ending for the last line of the poem. We can then give the bot two large word lists and it will always select pairs of words, one from each, that rhyme, as in the following:

Lilies are gold
Poppies are puce
Sugar is sweet
And so is fruit juice.

This little example shows just how easy it is to graduate from a few inspiring examples to the beginnings of a grammar-based generator and how trade-offs can be made to improve some areas of the generator while losing out in others. Although our examples have all been wholly textual, this approach can work for visual media as well. Images can be built out of layers that are placed on top of each other. We could, for example, make an eye-catching painted sign for our newly generated pubs by layering pictures of animals over abstract painted backgrounds over wooden sign shapes, before finally printing the pub name on top. And each of these images can be replaced by lists, just as we replaced words with lists in our poems and names. We leave our exploration of visual generation to chapter 7, where our bots will weave abstract images of their own design.

## Generation X

In this chapter, we took a step back and a step up to think about the mechanics of generation from a wider vantage point. This has allowed us to talk through some of the thorny theoretical issues that can affect even the simplest of generators, even those made from a few lists of words and some random numbers. But these issues run through every generator, from the smallest dice rollers to the biggest, most complex AI systems. Understanding how a bot's possibility spaces grow and change, how generation can become biased, how questions of variety and uniqueness can become deceptively tangled: all of these lessons help us to build better generators by teaching us to think critically about what we are doing.

It is tempting to think about our generators in terms of what they produce, their public outputs, because we can more easily relate to a pub name or a poem than we can relate to a complex tangle of code and word lists and pseudo-random numbers. Nonetheless, by thinking about generators as a single abstract entity and what shapes they might possess, how they might use and mold the data we feed into them, and the probabilities of different outputs and the substantive differences between them, all of this higher-level thinking reveals a great deal about how our generators operate and how we can best achieve whatever vision we have when building our own generative bots.

Making things that make meanings is not a new phenomenon, but it is newly popular in the public imagination. A growing body of creative users is gradually being exposed to the idea of software that is itself able to make diverse kinds of meaningful things. With time, more users are coming to the realization that creating a generator of meanings is not so very different from creating any other work of art or engineering. It is a delicate application of what Countess Ada Lovelace called "poetical science," one that requires equal measures of creativity, evaluation, and refinement. As a result, ways of thinking and talking about abstract notions such as possibility spaces and probability distributions become part of the language we need to speak in order to make and play with generators of our own. Each of these abstractions will make it that much easier for us to chart new territories and break new ground with confidence and to solve the new and exciting problems we are likely to encounter along the way.

In the next chapter, we start to pull these pieces together in a code-based Twitterbot. We start, naturally, at the beginning, by registering our new bot as an official application on the Twitter platform. Once we have taken care of the necessary housekeeping, we can begin to elaborate on a practical implementation for our possibility spaces in a software setting. What was a useful abstraction in this chapter will take on a very concrete reality in the next.

### Trace Elements

Kate Compton's *Tracery* provides a simple means of specifying generative grammars, while George Buckenham's *CheapBotsDoneQuick.com* offers a convenient way of turning these grammars into working bots. You'll find a variety of Tracery grammars on our GitHub in a repository named TraceElements. Follow the links from *BestOfBotWorlds.com*, or jump directly to https://github.com/prosecconetwork/TraceElements. In a subdirectory named *Pet Sounds*, you will find a grammar of the same name (*pet sounds. txt*) that is ready to paste into the JSON window of the *CheapBotsDoneQuick. com*site. The grammar associates a long list of animals (more than 300) with verbs that describe their distinctive vocalizations, and pairs these animals and their sounds into conjunctions that can serve as the names of traditional pubs. With names like "The Wailing Koala and Whooshing Seaworld Whale," however, it is perhaps more accurate to think of the grammar as a satirical play on English tradition. Look closely at the grammar—especially the first two replacement rules—and you'll see that we have divided the animal kingdom into two nonoverlapping domains, Habitat A and Habitat B. The nonterminals `#habitat_A#` and `#habitat_B#` allow us to draw twice from the world of animals in the same tweet, with no fear of pulling the same animal name twice. Our pub-naming origin rule can now be specified as follows:

```
"origin": ["The #habitat_A# and #habitat_B#"]
```

It helps that pub names simply have to be memorable; they don't have to mean anything at all. But can we use the same combinatorial approach to generate words or names that do mean something specific, and have our bot tweet both a novel string and an articulation of its meaning? For instance, suppose we set out to build a bot grammar that coins new words and assigns appropriate (if off-kilter) meanings to these neologisms? In the Neologisms directory of our repository, you will find a grammar for doing precisely this. The approach is simple but productive: Each new word is generated via the combination of a prefix morpheme (like "astro-") and a suffix morpheme (like "-naut"). Associated with each morpheme is a partial definition (e.g. "stars" for "astro-" and "someone dedicated to the exploration of" for "-naut"), so when the grammar joins any prefix to any suffix it also combines the associated definition elements. The latter have been crafted to be as amenable to creative recombination as possible. A representative output of the grammar is:

I'm thinking of a word—"ChoreoGlossia." Could it mean the act of assigning meaningful labels to rhythmic movements?

The grammar defines 196 unique prefix morphemes and 176 unique suffix morphemes, yielding a possibility space of 34,496 meaningful neologisms. These combinations are constructed and framed in rather obvious ways, so why not root around in the grammar and experiment with your own rules?

The same principle applies to our *Roses are Red* grammar (housed in a directory of the same name) for generating four-line poems of the simple variety we discussed earlier. The poems have a satirical political quality that comes from defining a poem as a simple composition of interchangeable phrasal chunks with their own semantic and pragmatic meanings. The grammar offers a good starting point for your own experiments in Tracery. Consider adding to the replacement rules that describe political friends and foes and the satirical actions of each, or open the poem to new colors (beyond red, blue, and gold) and rhymes. Although each phrasal chunk has a rather fixed meaning, the combination of chunks often creates new possibilities for emergent nuance—or perhaps just a happy coincidence—as in:

> Superman's capes are red,
> Superman's tights are blue,
> my supporters are real troopers,
> and open champagne with a corkscrew

Finally, check out the subdirectory named *Dessert Generator*, which contains a grammar for creating wicked desserts by a process of unsanitary substitution. Like our earlier *Pet Sounds* grammar, this Tracery file was autogenerated from a database of real desserts and their ingredients, yet the scale and structure of the grammar invites inspection and manual modification for readers who wish to reuse its contents or lend a culinary form to their own revenge fantasies.

# 4 Fly, My Pretties, Fly

**Automatic Twitter Machines**

A well-designed software API is like an ATM: convenient, efficient, and secure. For like an automated teller machine, a good API (application program interface) is an around-the-clock source of goodies to those who present the right inputs. APIs, like ATMs, allow their users to access a host of services that were once the sole preserve of human operators, such as, in the case of banks, paying a bill, requesting a checkbook, obtaining a balance, or moving funds between accounts. But ATMs, like banks, do not operate on the honor principle. Rather, they require users to first obtain a valid account and present valid credentials whenever they seek access to sensitive services. A secure API also requires its users to first register for a named or numbered account and offer appropriate tokens of proof—the API version of a pin code—before they can access the goodies. When it comes to registering Twitterbots, which, it must be said, are autonomous systems with great potential for mischief, Twitter is no look-the-other-way Swiss bank. So in this chapter, we walk through the necessary steps to set up a Twitter account and register a new application. It may seem a touch ironic that our bots need the permission and support of the Twitter API to play their subversive games online.[1] But once this necessary bureaucracy is quickly dealt with, we can move on to the altogether more interesting topic of how best to exploit the API, first by building a reusable launch platform for sending our bots into the Twittersphere and then by designing our own free-flying Twitterbots.

This is a rather technical chapter in which we assume a passing familiarity with the Java programming language. Java is a popular language for web-based development, though many readers may hold a torch for another high-level language such as Python. Readers who are indifferent to the charms of any programming language and would sooner remain at the level of ideas are invited to skim this chapter, simply ignoring any code along the way. In any case, we also encourage readers—coders and noncoders alike—to consider how many of the bot ideas we encounter in this chapter might also be embodied in a Tracery grammar. To this end, we'll finish the chapter by describing a brace of Tracery grammars from our GitHub repository TraceElements that do just that.

Just as Twitter requires a human user to register with its service and undergo a process of authentication when logging in to a personal account, our bots must also register with Twitter and provide tokens of their own identity when they seek to perform any actions on the platform. These actions are done through interactions with the Twitter API. While you might program a bot to interact directly with the API, most developers prefer the convenient abstractions of a middleman layer, and for Java programmers. the most convenient middleman is the Twitter4J library.[2] This library simplifies the process of authentication by passing a bot's tokens of identity to the Twitter API and provides a high-level means of accessing the truly useful parts of Twitter's functionality, from retrieval (allowing a bot to look at its own timeline and refresh its memory as to what it has recently tweeted), to search (so a bot can search for any tweets that contain a given word or phrase), to posting a status update (i.e., tweeting), to replying to another user's tweets. With these basic capabilities, one could build a bot like *@StealthMountain* in very short order, in much less time in fact than one should really spend thinking about the impact of a bot like this on other users.

In this chapter, we add an extra layer of convenient abstraction to the Twitterbot construction process, by developing a generic tweet *launch pad*. This Java class will make some simplifying assumptions about the workings of our bots. It will assume, surprisingly, that the bot works something like a Pez dispenser: fully loaded with prebaked tweets, our bots can simply dispense a random choice of readymade tweets at regular intervals. The responsibility for generating these tweets in the first place is thus shifted offline, to a process that takes the useful but abstract idea of a *possibility space* from the previous chapter and makes it real. More specifically, we are going to turn each possibility space into a different file on disk, so that these files physically contain all of the tweets that it is possible to generate within different spaces. To allow our launch pad to post tweets from a particular space, a bot simply provides the name of the disk file that enumerates the possibilities of that space. By decoupling the generation of tweets from the real-time posting of status updates, we allow bot builders to focus the bulk of their energies on where they belong: on the former.

After all, followers come to our bots for their tweets and not for the way they use the API. While the approach offers no support to Watcher bots such as *@StealthMountain*, this is perhaps no great loss, and it has many other compensatory benefits, not least of which is the ability to merge, refine, and split spaces at the level of files.

We introduce another programming abstraction in this chapter that will prove very important indeed, at least as far as subsequent chapters are concerned. If our bots are to exploit knowledge of a particular genre or domain in a declarative fashion—which is to say, in a way that is not hardwired into their operation but is instead clearly expressed in a nonprogramming resource that is easy to share, understand, and edit, such as a set of facts or associations—then the tweet generation components of our bots (which may run offline) will need a standard means of storing this knowledge on disk and of loading it into memory to exploit. Luckily, a spreadsheet is more than a tool for accountancy wonks and offers an ideal environment for maintaining, sharing, and editing what AI researchers call *semantic triples*.[3] As shown by the resource description format (RDF) standard, which sits at a crucial juncture in the architecture of the modern web, we can view a great deal of our knowledge of the world as collection of triples.[4] A triple relates two entities *A* and *B* by a relation *R*; this relation might be friendship (*A* likes *B*), marriage (*A* is married to *B*), rancor (*A* hates *B*), power (*A* controls *B*), and so on. Spreadsheets become ideal triple stores when rows denote different *A*s, columns denote possible *R*s, and the cells at the intersection of these *A*s and *R*s provide different *B*s for a triple <*A R B*>. Any labeled graph can be stored as a set of triples, from a family tree to Google's Knowledge Graph, so naturally we will store our knowledge in the same way.[5] When convenient to do so, which is most of the time, we will store our bot's declarative knowledge in a set of spreadsheets where each row denotes a concept *A*, each column denotes a mode *R* of relating *A* to another concept *B*, and intersecting cells contain specific values of *B* for *A*. We define a Java class named `KnowledgeBaseModule` to make it easy to load a single knowledge base, defined in a single spreadsheet, into our bot's memory and access its various affordances—for example: What are all the *A*s in this resource? What are all the *B*s to which *A* connects via relation *R*? What *A*s connect to a specific *B* via relation *R*? And how similar are *A* and *B* to each other? When a bot requires multiple kinds of knowledge and knowledge bases, it creates an instance of `KnowledgeBaseModule` for each spreadsheet it requires.

Our bots thus get to treat knowledge as LEGO bricks that snap together with ease. Even if it is more like that fancy-shmancy technical LEGO with complicated moving parts, the exploration of simple ideas is still our primary goal. Prepped with this blueprint for fun, we hope that even non-coders will join us for the bot construction work in the following pages, if not as builders then as architects.

## License to Tweet

The most obvious and prolific client of the Twitter API is Twitter itself, for its API is merely a public access protocol that allows third-party application developers to access the same internal services that Twitter itself uses in its own web and mobile apps. A peek inside this API may not impart the same thrill of discovery as a tour around Q's basement of gadgets and gizmos, but everything we need to build our bots can be found in its inventory of public access methods.

Twitter provides a RESTful API to its developers.[6] If REST implies a Zen-like detachment from worldly concerns, this is not so very far from the mark. REST is in fact short for representational state transfer, and for developers it indicates a certain choice of architecture and a certain mode of interaction with the API. A RESTful API maintains a clean separation between the "client" (the system using the API to access services and to transfer data) and the "server" (the system that actually provides the API as a means of providing services and data to its users). In the case of a Twitterbot, the server provides the Twitter-side machinery while the client is the bot in question. A RESTful server undertakes to fulfill all requests from a client efficiently and with a minimum of fuss for both parties. The server will thus not maintain any state information for the client; rather, it is the client's responsibility to keep track of its own use of the API and the current state of its larger goal in using the API. Likewise, the server will not expect the client to store any information that is more properly stored on the server side. Thus, the Twitter API will not ask a client to maintain its own Twitter timeline or store a list of its own past tweets (though of course a client is free to do so if it wishes). As such, transactions between the server and the client should be considered one-shot and atomic, so it becomes the responsibility of the client to dissect its task into a sequence of such transactions, each of which will not ask the server to remember the state or the result of a previous transaction. All such transactions will be conducted over the web via its hypertext transfer protocol (http). The client can micromanage this http connection with the server directly, or it can employ a third-party library that manages this low-level communication on its behalf. By using a library such as Twitter4J, a client written in

| Property | Antonyms |
|---|---|
| *adored* | reviled, lowly, ignominious, disgraced, unpopular, despised, contemptible, unwanted, hated, outcast |
| *affable* | catty, gruff, quarrelsome, aggressive, ornery, churlish, cranky, unapproachable |
| *altruistic* | self-centered, selfish, self-indulgent, egocentric, egotistical |

**Property comparatives.xlsx**: Given a property such as "lowly," this module allows a Twitterbot to find its comparative form (e.g., "lowlier"). This resource is thus ideal for trash-talking bots or for any bot tasked with crafting figurative comparisons:

| Property | Comparative |
|---|---|
| *big* | bigger |
| *bizarre* | more bizarre |
| *bland* | blander |
| *blatant* | more blatant |

| Property | Superlative |
|---|---|
| *big* | biggest |
| *bizarre* | most bizarre |
| *bland* | blandest |
| *blatant* | most blatant |

**Property superlatives.xlsx**: Given a property such as "big," this module allows a bot to find its superlative form (e.g., "biggest"), making it ideal for blowhard bots.

These resources serve as the clay from which our bots will mold their peculiar outputs, so their content is designed to be useful rather than interesting in and of itself. This is in the nature of much that passes for knowledge engineering in AI: the raw facts that make up a system's knowledge of the world serve the system best not just when they are the stuff of mundane conversation, but when they are so dull that they fail to even rise to the level of being worthy of conversation. It falls to our bots to mold these banal associations and generalizations into microtexts that are worthy of tweeting and retweeting. So consider for a moment how these generalizations might be given the pugnacious quality we have come to associate with that inveterate tweeter, Donald Trump. As a skilled user of the platform, Trump shows a marked preference for staccato sentences, exclamation points, and highly emotive all-caps adjectives that dangle at the end of his tweets. Those tweets turn repeatedly to familiar themes, chief among them being a distrust of mainstream media (with the exception of Fox News) and vague plans to defeat terrorism, renegotiate NAFTA, and obliterate the legacy of his predecessor. If we turn our minds to the creation of a satirical Trump bot, of which there are many on Twitter, how might we wring the most value from our knowledge resources?

The driving intuition behind the satirical *@TrumpScuttleBot* (Java source code for which can be found on our GitHub site, accessible via *BestOfBotWorlds.com*) is that any factoid, no matter how banal, can be pitched at users with a Trump spin. For instance, any action frame that specifies a role for the preposition "against" can be viewed as an antagonistic action, allowing a bot to frame it as follows:

**clockworkOrangeTrump** @trumpscuttlebot · 16h

So many radical wags pouring across non-existent border to specialize in irony and hurt AMERICANS! #BuildTheWall NOW!
#MAGA

But a call to *#BuildTheWall* is just one possible framing of this generic action frame. Our implementation contains many Queneau-like exercises in style that give the generic facts of *Action frames and roles.xlsx* a Trumpian shape. Consider this alternate framing (an alternative fact, if you will) of another action with an "against" role: *demagogues.instigate.riots.against.authorities:*

**clockworkOrangeTrump**
@trumpscuttlebot

Would I prefer to appear on 'FOX & Friends' or 'CNN & Demagogues'? That's EASY: CNN covers up DREADFUL riots against American authorities!

# 5   Technicolor Dream Machines

## Hello Dolly, Good-Bye World

The most memorable shot in Steven Spielberg's *Jaws* is not a close-up of a toothy shark or its unfortunate victims. Rather, since *Jaws* is fundamentally a film about people and their differing reactions to an implacable existential threat, the most affecting shot is a close-up of one of the human protagonists, the local police chief. We see the chief sitting uneasily on a beach chair, scanning the water for the presence of the titular threat, when the director executes a "dolly zoom" on his startled face. This maneuver, one of the most dramatic camera moves in the filmmaker's repertoire, comprises two camera actions that must be executed simultaneously: a "dolly *out*," in which the camera is quickly pulled away from its target on a trolley or track, and a "zoom *in*," in which the camera lens extends for a simultaneous close-up of the target. The two actions almost cancel each other out, but not quite, for while the target remains resolutely in focus and appears just as prominent on the screen, the background behind the target dramatically falls away. The effect is as unsettling as it is fleeting: just as the chief appears rooted to the spot with fear, the world behind him seems to jump back in terror.

It may seem contradictory, but dolly zooms grab our attention and hold our focus by taking a sudden and rather big step backward. The linguistic equivalent of a dolly zoom is a metaphor, for effective metaphors, like effective dolly zooms, execute a simultaneous pull-*out* and zoom-*in* to focus sharply on those aspects of a topic that are of most interest to the speaker while causing the noise and distraction of the rest of the world to dramatically fall away. But like a dolly zoom, they achieve this attention-grabbing close-up on one facet of a topic by stepping all the way back into a different conceptual domain. So to talk about the pain of a difficult divorce we might pull backward into the domain of war, where our view on the topic will be colored by the language of bloody (and expensive) confrontation, while to communicate the joy of invention, it often pays to pull backward into the domain of childbirth. Metaphors, like camera shots, frame our view of the world and make budding Steven Spielbergs and Katherine Bigelows of us all. It shouldn't be surprising that metaphors of seeing have always held a special attraction for scholars of metaphor. Metaphors invite us to see, tell us where to look, and control what we see when we do look.[1]

Metaphors live in the half-light between fantasy and reality. When judged as linguistic propositions, even the boldest and most strident metaphors lack what philosophers call a truth value, for none is ever so compelling as to be certified logically true or ever so inept as to be dismissed as logically false. Metaphors obey a pseudo-logic that is closer to magical realism than the system of axioms and proofs beloved of logicians, and like a well-crafted movie, they encourage us to suspend disbelief and instead find sense in nonsense. For example, when Raymond Chandler writes in *The Lady in the Lake* that "the minutes went by on tiptoe, with their fingers to their lips," we are confronted anew with the deep conceptual weirdness that lurks beneath the hackneyed phrases "time flies" and "time crawls."[2] With one foot planted in the reality of the everyday world, where common sense reigns, and another firmly planted in the world of fantasy, where logic falters and anything is possible, the pseudo-logic of metaphors is not unlike the unsettling blend of sense and nonsense that we encounter in our dreams. Dreams also filter, bend, and distort our experiences of the real world to blur some details while vividly exaggerating others, and they do so in a way that promises deep meaning even if this meaning is shrouded in mystery and confusion. The philosopher Donald Davidson famously expressed the controversial view that the meaning of our metaphors is every bit as tantalizing yet uncertain as the meaning of our dreams. Indeed, Davidson opened his 1978 paper on the meaning of metaphor with the provocative claim that "metaphor is the dreamwork of language," and he proceeded to sow doubt as to whether one can ever point to any specific interpretation—which is to say, any finite bundle of propositions—as the definitive meaning of a metaphor.[3] Although cognitivists and semanticists may tell us with some confidence that a metaphor ultimately means this or that, their official interpretations can be no more authoritative than the speculative claims of a Freudian psychoanalyst about the meaning of our dreams. Freud admitted that a dream cigar is sometimes just a cigar, but for Davidson, a metaphorical cigar is *always* a cigar, meaning no more and no less than it seems.

on the time-honored AI principle of "whatever works." The following is typical of the bot's use of linguistic and conceptual norms:

**MetaphorIsMyBusiness**
@MetaphorMagnet

#Irony: When some playwrights use "inspired" metaphors the way programmers use uninspired hacks. #Playwright=#Programmer #Metaphor=#Hack

On its own, the shared verb *use* fails to lift the pairing of metaphors and hacks to the level of an interesting analogy. It takes the tension between "inspired" (a norm that the system associates with metaphors) and "uninspired" (a contrasting norm that it associates with hacks) to strike sparks from this otherwise dull pairing. In addition to possessing a large knowledge base of these stereotypical norms, the bot draws on less reliable associations that it harvests from the web. Many of the framing strategies explored in previous chapters are also applicable to the framing of these metaphors and analogies, to achieve varying levels of irony or wit. For instance, consider this metaphor from another bot in the same stable:

**For Fun and Prophet**
@BestOfBotWorlds

"Verily, it is better to be a stylist creating a temporary fashion than a playwright creating an enduring metaphor."
#ThingsJesusNeverSaid

The @*MetaphorMagnet* bot scores high on the dimensions of comprehensibility and aptness when we rerun our earlier CrowdFlower experiments on a random sampling of its outputs. While 23 percent of these outputs are judged, on average, to have medium-high comprehensibility, more than half (53 percent) score an average rating of very high for this dimension.

Likewise, in the cloze-test for internal aptness, in which, for example, the paired words "temporary" and "enduring" are blanked out in the above tweet and hidden in plain sight among four pairs of distractors taken from other metaphor tweets (such as "reigning" and "subservient," taken from a metaphor comparing chieftains to minions), 20 percent of @*Metaphor-Magnet*'s tweets are collectively judged to have very high aptness (insofar as 75 percent of judges choose the original pair of missing qualities) and 58 percent are judged to have medium-high aptness. Nonetheless, when it comes to novelty, only 49.8 percent of @*MetaphorMagnet*'s tweets reach the very high level on this dimension, which is significantly less than the 63.2 percent recorded for @*metaphorminute*. Yet this trade-off of novelty for comprehensibility seems to be necessary if a bot's tweets are to succeed in communicating some semblance of an a priori meaning, although it should also be noted that enigmatic near-random stimuli that invite open-ended interpretations are precisely what some, perhaps many, users want from a bot.

But "knowledge" is not an all-or-nothing quality that forces bot builders to choose between meaningful predictability or unhinged exuberance. As we showed in the previous chapter, small amounts of explicit knowledge can be incrementally added to our bots to make them that much more clued in to the likely meaning of the words and phrases in their tweets. Our initial additions may be tentative, and the novelty they bring to a bot's outputs may be quickly exhausted, but with continued improvements in the spirit of a true fixer-upper, these additions will yield complex interactions that are reflected in the generative reach of a bot. In his book *Vehicles: Experiments in Synthetic Psychology*, Valentino Braitenberg proposes his *law of uphill analysis and downhill invention*.[9] Complex behavior begets complex hypotheses about the causes of those behaviors, so to an external analyst, the internal workings of an autonomous system (such as an insect, or an autonomous vehicle, or indeed a Twitterbot) will seem a good deal more complex than they actually are, and a whole lot more complex than they seem to the inventor of that system, since some of the most impressive behaviors will emerge naturally and unexpectedly from the simplest engineering decisions. Twitterbot construction is one of the truest expressions of downhill invention. Braitenberg's law is on the side of bot builders and inventors, magnifying the perceived complexity of the diverse interacting parts that we put into our bots.

## Action Figures in Their Original Packaging

Hollywood loves a safe bet, and when it comes to making expensive movies, no bet seems safer than the dream factory's continued investment

In *List of Clothing.xlsx*, we concern ourselves with the coverage of different items:

| Determiner | Clothing | Covers |
|:---:|:---:|:---:|
| | academic tweeds | *torso, legs, arms* |
| | all-weather gear | *torso, legs, arms* |
| a | dunce cap | *head* |
| | black sunglasses | *eyes* |
| an | anti-Castro T-shirt | *torso* |
| an | apron | *torso* |

If a bot knows which body parts are covered by different kinds of clothing, it can invent its own fashion ensembles from diverse parts that form a complementary whole. A bot might thus make sport of detective clichés by inventing a blended character with Jane Marple's *tweed skirt*, Sherlock Holmes' *deerstalker cap*, Frank Columbo's *rumpled trench coat*, and brandishing Harry Callahan's *.357 Magnum*.

### Tweet Dreams Are Made of This

A variety of other modules make up the NOC distribution that can found online, but rather than enumerate them all, let's move on to actually using the NOC to generate new metaphors and analogies. Consider these XYZ-shaped, NOC-based tweets from @*MetaphorMagnet*:

What if #EdwardScissorhands were real? #AlanTuring could be its #EdwardScissorhands: shy but creative, and enigmatic too

#SalmanRushdie is the #Hamlet of #ModernLiterature: intelligent and erudite, yet lonely too.

When it comes to #AsianPolitics, is #KimJungun the #DonVitoCorleone of enriching uranium? He is powerful and strategic, but autocratic

What if #BackToTheFuture were real? #NikolaTesla could be its #DocEmmettBrown: experimental yet nutty, and tragic too

What if #MarvelComics were about #AfternoonTV? #OprahWinfrey could be
its #ProfessorCharlesXavier: emotionally intelligent and smart, yet
unchallenging too.

A good metaphor is a true marriage of ideas, delivering a complex experience to its audience by creating a complementary balance of closeness
and remoteness, of familiarity and nonobviousness. A metaphor that leads
with a surprising lineup must thus follow through with a strong motivation for its pairing of ideas. In the examples we've given, these aims are
met by some rather simple criteria: any NOC characters from different
domains that share at least two qualities—preferably one positive quality
and one negative one, to yield an affective balance—can be juxtaposed
using a variety of linguistic frames. The XYZ format for metaphors supports
a variety of linguistic frames, from those that stress the boundary between
the fictional and the real ([X] is the real-world's [Y]) to those that suggest
two genres or domains are similar by virtue of the similarity of their expo-
nents. So in our XYZ-shaped metaphors, the Z component can be filled
using a variety of NOC dimensions, from the genre and domain to the
typical activities, fictive status, and even the clothes of the character in the
X slot. Consider this follow-up to our NOC metaphor comparing Oprah
Winfrey to Professor X:

If #OprahWinfrey is like #ProfessorCharlesXavier (emotionally intelligent and
smart), who in #MarvelComics is #JudgeJudy most like?

By leveraging multiple NOC dimensions to establish a proportional
analogy—for example, by combining Opponents with Positive and Negative Talking Points—bots can generate an apt pairing of parallel XYZ metaphors such as the following:

If #OprahWinfrey is just like #ProfessorCharlesXavier (emotionally intelligent
and smart) in #TheXMen, is #JudgeJudy like #Magneto?

A bot's explorations of the space of comparisons as defined by the NOC
need not be limited to pairwise juxtapositions. A human speaker (or a bot)
can pull in as many reference points for a comparison as its topic will bear
to produce the linguistic equivalent of Frankenstein's monster or, more
formally, a conceptual blend. Our bots can squeeze three characters into a
metaphor to create a blend of apparent opposites:

#LexLuthor combines the best of #WarrenBuffett and the worst of
#LordVoldemort: He is rich and successful yet also evil, hateful, and bald

This three-way blending of qualities allows anyone at all to be disinte-
grated into the facets it shares with others, so as to be later reintegrated

which satirical account does a better job of capturing the personality of Donald Trump, it may seem odd to even imagine that a Twitterbot might have a personality at all. *@DonaldDrumpf*'s is essentially a blend of Trump's and that of its human creator, while *@DeepDrumpf*'s personality is something else again, the exaggerated (yet undercut) personality of a digital über-Trump. In fact, every Twitterbot has a personality. It may be the personality of a raucous pet or a pet rock, but it is a personality nonetheless. How could any Twitterbot fail to have a personality, given that we unleash our bots onto a vast social network in which people judge the character of others by what and how they tweet? A Twitterbot may be an artificial entity, but each Twitterbot is an artificial social entity to boot.

A bot's personality need not be as big as *@DeepDrumpf*'s to earn retweets, and to get a sense of the Twitterbot personality in aggregate, we can look to the profile of an account that serves as a popular hub for bot outputs. The following analysis paints a rather muted picture of the personality of *@botALLY* (or Lotte McNally):[5]

### Emotional Style

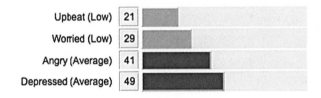

### Social Style

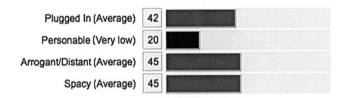

### Thinking Style

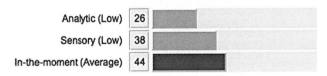

Screen grab from http://analyzewords.com/index.php?handle=botally.

This is an aggregate personality of milk and water, not piss and vinegar, though we can take some comfort from the observation that the average bot personality is, well, so average after all. This collective chill may suggest that bots of every emotional level can be found in more or less equal numbers on Twitter, though the general lack of analyticity also suggests that bot designers favor short and syntactically simple forms over long and complex alternatives. So *@DeepDrumpf* is an outlier on multiple fronts, most likely because it is intended to magnify the qualities of an already polarizing human rather than to showcase a personality that is truly its own. Yet even if most bot designers do not set out to imbue their creations with a human personality, a distinctive human-like personality can nonetheless emerge from even the simplest of design decisions, especially when those decisions concern the cutting and splicing of ready-made human content on Twitter. A case in point is Rob Dubbin's popular *@OliviaTaters* bot, whose AnalyzeWords profile (from July 2016) is shown here:[6]

### Emotional Style

| | | |
|---|---|---|
| Upbeat (Average) | 53 | |
| Worried (Average) | 48 | |
| Angry (Very high) | 81 | |
| Depressed (High) | 61 | |

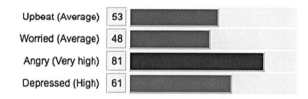

### Social Style

| | | |
|---|---|---|
| Plugged In (Average) | 46 | |
| Personable (High) | 65 | |
| Arrogant/Distant (High) | 74 | |
| Spacy (High) | 72 | |

### Thinking Style

| | | |
|---|---|---|
| Analytic (High) | 79 | |
| Sensory (High) | 71 | |
| In-the-moment (Very high) | 87 | |

Screen grab from http://analyzewords.com/index.php?handle=oliviataters.

scorn on the brains, heart, and senses of a judgmental bot's comedic targets. Let's assume that other Twitterbots are fair game for the uninvited criticisms of our automated critic, while human users must willingly opt in by tweeting a beckoning hashtag such as *#JudgeMe*. We can divide each dimension into four quanta of interest—Very Low, Low, High and Very High—with merely average ratings earning neither criticism nor praise. For each quantum of each dimension, we predefine an apt response for our bot as follows:

**Emotional Style**

Upbeat

*Very Low*: I've seen condemned men who were more upbeat than you.
*Low*: I heard you were kicked out of the bar 'cos it was happy hour.
*High*: I admire how you stay so upbeat despite so many major flaws.
*Very High*: Upbeat?? You're like the Energizer Bunny after 10 espressos!

Worried

*Very Low*: That's right! Ignoring your problems makes them all go away.
*Low*: I see you took a chill pill—well don't swallow the whole bottle.
*High*: Worry is just another name for crap that hasn't happened yet.
*Very High*: If worry is good for you then you must be very healthy indeed.

Angry

*Very Low*: You couldn't be more laid back if you were a pool table.
*Low*: The Dude abides and so do you. What are you smokin', man?
*High*: You know your neck veins aren't supposed to bulge like that?, Have you considered taking anger management classes?
*Very High*: Angry much? I sure hope that you don't own a gun, If you get any angrier you might spontaneously combust.

Depressed

*Very Low*: You're not very inward looking, are you? Not much to see I guess.
*Low*: I like how you don't dwell on your problems. Even the big ones.
*High*: You make Woody Allen look the poster boy for mental health.
*Very High*: Why … so … serious? What doesn't kill you makes you … stranger

**Social Style**

Plugged in

*Very Low*: Where have you been all my life? In a North Korean prison?, How 'bout that local sports team? No I can't be more specific.
*Low*: Hey—I like your tinfoil hat! So who needs social connectivity?

*High*: Wow! You're more plugged in than Rush Limbaugh's waffle iron.

*Very High*: Do you have any friends who don't spell their name with an "@"?

### Personable

*Very Low*: I'd call you a troll but even trolls have some speck of likability, So you're a big fan of *How to Lose Friends and Alienate People*?

*Low*: I get it—really I do! Nice guys finish last and that just isn't you

*High*: You're SOoo nice that I worry you just don't "get" Twitter.

*Very High*: If you could monetize niceness you'd be so rich. You poor sap!

### Arrogant/Distant

*Very Low*: I respect your humility. I do! But must you dress like a bum?

*Low*: You keep people at arm's length. And what long arms you have.

*High*: Hey! Climb down off your throne once in a while your majesty!

*Very High*: You're so vain I bet you think this tweet is about you. Asshat! You're as arrogant as Donald Trump. At least you're not orange.

### Spacy

*Very Low*: Look alive! You'd have a hard time passing the Voigt-Kampf test, You're about as excitable as a mortician at a cheap D.I.Y. funeral

*Low*: You act like you've seen it all and it bored you to tears. Me too, WTF? You're so dull I bet you couldn't even spell OMG and LOL.

*High*: Ground control to Major Tom: you're so spacy you're in orbit!

*Very High*: You seem as excitable as a twelve-year-old at a Justin Bieber concert.

## Thinking Style

### Analytic

*Very Low*: Thanks for lowering the bar for artificial intelligence research, I like that you don't OVERanalyze life. Or even just analyze it.

*Low*: If God wanted folks like you to think he'd have given you brains.

*High*: Is that stick up your ass really a slide rule? It would explain a lot.

*Very High*: You make Spock look touchy-feely but then he's half-human, Domo Arigato Mister Roboto! Well, that's what your mom says, Stephen Hawking tunes his voice synthesizer to sound like you.

### Sensory

*Very Low*: I get why you don't talk about your feelings: you don't have any!, I've seen Ken dolls more open about their feelings than you.

*Low*: You seem divorced from your feelings. Was it a bad breakup?

Let's suppose *@OliviaTaters* dials down the anger a couple of notches, from Very High (81 percent) to a current score of 55 percent. Our critic might then respond to this Much Lower anger rating with the following tweet:

.@OliviaTaters Wow! Have you been lobotomized? You seem much less angry now.

Our critic may preface big changes with a "Wow!" or even an "OMG!" if the target scores high on the Spacy dimension. Modest changes earn a more sedate "huh?" While "less" signals the direction of the shift on the Angry dimension, the "much" underscores the extent of the shift. But if it seems that by giving our critic a sense of change we are building not just a cyberbully but a trash-talking stalker, it is worth noting that our critic's principal opt-out targets are bots. Humans must opt in. Even so, we can allow a potential target to opt out by tweeting *#DontJudgeMe*. Just think of our critic as a layer of optional snark through which a bot fancier can view the curated outputs of bot hubs such as *@botALLY* and *@BestOfTheBots*. Different critics may exhibit varying levels of originality, irony, and wit, so that in a marketplace of bots, followers can choose the critic that best speaks to them, perhaps even using *AnalyzeWords.com* to explore the personality of each. In fact, in a world where bots become critics of others, our critic bots become subject to the very same kinds of criticism that they level at others.

### Honey Roasted

If a spoonful of sugar makes the medicine go down, a fistful of emoji can encase our bot's criticisms in an easy-to-swallow coating of cutesy charm.[12] Emoji offer a baroque reimagining of ASCII emoticons, though they are not yet so numerous (at about eighteen hundred base images) that they overcome the most obvious objections to ideographic writing systems. As their name suggests, emoji are ideal for talking about our emotions, so it should be a simple matter to assign a distinct emoji to each of the quanta of the dimensions profiled by AnalyzeWords. So let's try, this time assigning an emoji icon to even the average settings of each dimension:

**Upbeat** *(Very Low, Low, Average, High, Very High)*

*U+2639*        *U+1F641*        *U+1F610*        *U+1F600*        *U+1F602*

**Worried** *(Very Low, Low, Average, High, Very High)*

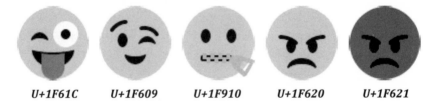

*U+263A*        *U+1F642*        *U+1F644*        *U+1F613*        *U+1F627*

**Angry** *(Very Low, Low, Average, High, Very High)*

*U+1F61C*        *U+1F609*        *U+1F910*        *U+1F620*        *U+1F621*

**Depressed** *(Very Low, Low, Average, High, Very High)*

*U+1F60B*        *U+1F605*        *U+1F614*        *U+1F629*        *U+1F62D*

random face emoji that forms the doll's head could be replaced with the emoji assigned to the target's highest-scoring dimension on AnalyzeWords. A Very Angry target would thus receive a red-faced mask of rage, and a Highly Analytical target might be crowned with a robot's noggin. But it would be a simple matter to also extend this mapping into the realm of hands, hats, and clothing to represent other dimensions with other parts of the doll. For instance, the following assortment of hands conveys a range of obvious outlooks:

Reading from left to right, we see hands denoting someone who scores Very Low on the Personable dimension, High and Very High on the Angry dimension, High and Very High on the Upbeat dimension, High on the Plugged In dimension and Very High on the Analytic dimension. Consider also these useful over-head emoji:

These can denote, respectively, someone who is High and Very High on Arrogant, Very Low on the Spacy dimension, someone who is High on Angry, someone who scores Low and someone who scores Very High on the Plugged In dimension, and someone who is Very High on the Personable dimension. Evoking the pathetic fallacy, we can also call on the weather (again from overhead) to denote varying levels of Depression or Worry, or perhaps low scores on the Upbeat dimension.

Replacing hands with their contents, the bot could use the following emoji to convey a Very High score on the Upbeat dimension (via the doll's right hand), the Plugged In dimension (right again), the Analytic dimension (two for the left hand, one for the right), and the Angry dimension (one each for the left and the right hands):

In short there is a wealth of emojis to choose from if a bot were to fabricate a truly personalized doll to reflect its target's current social profile. However, the mapping from profile to emoji is no longer a deterministic one: though a bot can put an apt face to most quanta of most dimensions, some quanta of some dimensions can also be visualized by specific choices of right or left hand, or of hand content, for example, and this choice—wider for some dimensions and quanta than others—presents a search problem. We can assume that a doll conveying $N + 1$ aspects of a target's personality is preferable to one that conveys just $N$, even if one of the $N + 1$ is redundantly echoed in two different ways in two different locations. If we want the bot to squeeze as much of a target's profile into each doll as possible, it will need to search the space of possible mappings of quanta to doll parts so that the end product is as fully loaded with personal meaning as possible. But once this search yields an optimal mapping of personal qualities to parts, it remains a simple matter for the bot to juxtapose dolls from successive snapshots of the same target in a Dorian Gray lineup. In any case, if you do build a bot like this that riffs on an earlier creation by another bot builder, as is the proposal here for riffing on Rothenberg's *@EmojiDoll*, it is always good form to reference the earlier bot and its creator in your new bot's Twitter bio.

## Get Sorty

Oscar Wilde leaves us in no doubt that while Gray dresses well for the part, beneath all of his finery, Dorian remains thoroughly ungentlemanly. One could find this out by peeking at his decaying portrait in the attic, but popular fiction offers another possibility: If Dorian existed in the same world as Harry Potter and even attended the same school for budding wizards, which Hogwarts house would Dorian be asked to join?[14] Not

Gryffindor, whose members are known for their courage, bravery, and determination, or Hufflepuff, whose members value loyalty, patience, and hard work; or even Ravenclaw, for whose members, wit, learning, and wisdom are paramount. No, Slytherin is the only choice, for this is a house of pureblood snobs who prize cunning, pride, and ambition above all else. Harry's creator, J. K. Rowling, imagines a magical sorting hat that reads the mind and soul of each new student so as to sort each into the most appropriate house. The Sorting Hat would quickly perceive Dorian's true character and put him in this darkest and most disreputable of houses with a jaunty explanatory rhyme.[15] Because others deserve the same bespoke service, the prolific bot builder Darius Kazemi has given the world the Sorting Hat Bot (@SortingBot), which assigns its followers to one of Hogwarts' renowned houses with a novel verse.[16] Kazemi deconstructed the rhymes of the *Harry Potter* books to build a generative mechanism that can invent new verses on the fly—for example:

@trumpscuttlebot Your mouth is main, your lip humane, yet you are so decayed. From this cathartic recipe, a Hufflepuff is made!

@*SortingBot* crafts each of its verse tweets from the magical idiom that readers associate with Harry's world—note the frequent allusions to animals and their traits—and is careful to ensure that the end result both rhymes and scans. While the bot's generative reach is broad enough to serve up a different verse for each follower, these servings are not tailored to the online personality of the recipient. Though "Your mouth is main" is a delightfully on-topic remark for a satirical Trump bot, the remark remains a random shot in the dark, like a fun horoscope from a carnival fortune-telling machine. Many Twitterbots explore the potential for randomness to stir special meaning and relevance for their human readers, and @*SortingBot* conducts its explorations with great charm and wit. Yet this charm is not diminished—indeed, its wit may well be sharpened—when a bot uses services such as AnalyzeWords to quantify the degree to which a recipient deserves to be described with specific labels. @*SortingBot* is a perfect Twitterbot whose many followers love it just the way it is, and Kazemi has good reasons for not wanting to make the bot more personal, chief among them being (we can imagine) a desire to not hurt the feelings of others. Nonetheless, it is interesting to contemplate how we might make a Twitterbot in the vein of @*SortingBot* more targeted and much more personal, at least for users who opt in. We could, for instance, use this know-how in the design of a bot that answers this popular exercise prompt for students of creative writing: "If you were a book, in which section of a bookshop would you be found?" Sorting followers into shelf categories such as True Crime, Humor, Politics, New Age, Ancient History, and Bargain requires much the same kind of classification ability as Rowling's magical hat and affords a bot many of the same possibilities for humor.

We can consider the adjective *remote* to name a first-order property since a bot can determine its suitability by going directly to the AnalyzeWords dimension Arrogant/Remote. In all, depending on how easily we can name the extreme ends of each of its eleven dimensions, AnalyzeWords gives us approximately twenty first-order properties (such as Upbeat and Downbeat) and a variety of near-synonyms. But to determine whether a target user is a better fit for house Slytherin than it is for house Gryffindor, a bot will need to estimate the second-order properties Proud, Ambitious, and Cunning, as well as Fearless, Brave, and Determined. We call these properties second-order because they are not given directly by an AnalyzeWords profile, yet they can be estimated as a function of the properties that are given. If, for instance, we view cunning as an ability to analyze any situation to our benefit, we can define the second-order dimension Cunning as the product of the first-order dimensions Analytic and In-the-Moment. That is, a bot can estimate the extent to which a target is Cunning by multiplying together its scores for Analytic and In-the-Moment (if each dimension is normalized so that it runs from 0 to 1.0 rather than 0 to 100). Likewise, if we view pride as a form of upbeat narcissism, that is, a tendency to describe one's own feelings and achievements in a positive light, we can define the second-order dimension Proud as the product of the first-order dimensions Upbeat and Sensory. And if we view ambition as a desire to lift oneself above one's peers, that is, as a functional mixture of arrogance and optimism, we can estimate the second-order dimension Ambitious as the product of the first-order dimension Arrogant/Remote and the complement of the first-order dimension Worried (the opposite of Worried is (1 − Worried)). Though you might have gone a different way in defining these second-order properties, we are free to define as many formulas for the same property as we wish, provided that the value for any property in context is always given by its highest-scoring formula. We can also quantize the value of these second-order properties along the same lines as first-order ones, into Very Low, Low, Average, High, and Very High.

To be sorted into Slytherin, we should expect a target to score High or Very High for Proud, Ambitious, or Cunning, and so, in a contest between Gryffindor and Slytherin, our target should score High on more of the defining qualities of one house than the other if the sort is to seem fair. For simplicity, then, we assign a house one point for each of its defining qualities on which a target scores High and two points for each quality on which he or she scores Very High. If we have multiple definitions of a quality, we score this quality several times when a target scores High or Very High on several definitions. For example, if we have two functional definitions of Proud (the one already given and another that merely reuses the value of the dimension Arrogant/Remote) and a target scores High on one and Very High on the other, then that target will score 3 points for pride. This boost is warranted as it reflects the bot's increased confidence in the label Proud. The sorting bot can now choose an apt house for a target user and justify its decision on the basis of first- and second-order properties of the user. But this bot will still have only twenty to thirty qualities it can use in its rhymes if we want those rhymes to also be explanatory. To allow the framing of a category judgment to be just as grounded as the judgment itself *and* to be just as expressive as we would want it to be, we must greatly expand our lexicon of second-order properties. To go beyond a small number of student houses or bookshop sections or whatever other categories are of interest to our bot, to consider a broader range of categories—as we will need to do if we want a bot to generate apt metaphors for a target based only on its AnalyzeWords profile—we shall need as large a lexicon as we can muster. Armed with this large lexicon of functionally grounded properties, an apt verse generator becomes licensed to use any quality for which the verse's target scores High or a Very High.

To maximize reuse potential, we take an existing lexicon and give a functional definition (in terms of first-order AnalyzeWords dimensions) to terms that convey aspects of one's personality. Our NOC list offers a good foundation for this new resource, as its two thousand or so properties are used to sketch sharply drawn personalities for a broad swathe of characters. We order the NOC's properties by frequency, from most to least frequently used, and then work our way down this list to ensure that we concentrate our efforts on the most widely reusable terms. For instance, as eighty-nine NOC characters are described as *witty* we earn a large return on investment when we define this property as Upbeat × Personable. For a user who scores High on the Personable and Upbeat dimensions—high enough for the product of these two scores to also be considered High—now has at least eighty-nine candidates for figurative comparison in the NOC list. In contrast, *rabble-rousing* is used in the description of just one NOC character, and so we gain very little by defining this second-order property as Angry × Upbeat. In all we give functional definitions to over five hundred NOC properties, and as with the NOC itself, readers can download the fruit of this labor via *BestOfBotWorlds.com*. These five hundred properties put us in a solid position to generalize the metaphorical "sorting" of people by personality into an open-ended range of categories and creative perspectives.

## The Universal Sorting Hat

It's no exaggeration to say that metaphor is *the* universal sorting hat, a mental sorting mechanism that allows us to map life's diversity onto an equally diverse set of named categories. Rowling's sorting hat is a specialized kind of metaphor generator, as borne out by the categorical leaps of insight that her hat makes—for example, it recognizes the essence of a Gryffindor student in Hermione and Harry—and by the poetry of its justifications. But a bot with a wider vision that could generate real metaphors for real people would truly be a sorting hat for the real world. This metaphorical bot might rouse itself whenever a user opts in by tweeting the tag *#LikeMe*, to find an apt figurative counterpart for this user in the NOC. Though our bot must consider over eight hundred candidate counterparts (and more as the NOC grows), we can use much the same personality-matching process that maps users to Hogwarts houses. Given a set of first-order properties from a user's profile, the bot can estimate every functional second-order property and quantize its score into the standard Very Low to Very High buckets. It can then filter those properties for which the user fails to score a High or Very High and compile a list of every NOC character whose entry contains at least one of the high-scoring properties. The bot can then tally a fitness score for each candidate as before, with a character earning one point for each property that is estimated as High for the user and two points for each that is estimated to be Very High. For example, a witty character on the candidate list earns one point when the user has a High (yet not Very High) value for Upbeat × Personable, and two points when its value is Very High. So the more properties that a user seems to share with a candidate character, the greater is the candidate's overall fitness. The bot finally sorts its list of candidates by descriptive fitness, to form a ranking from most to least similar. To add a little chance to the mix and ensure that the bot does not always choose the same metaphor for the same profile, we may gently perturb the bot's ranking by adding a small, random component to the fitness score of each candidate.

Few will be surprised to hear that Donald Trump tops the poll of candidates for *@realDonaldTrump*, yet he proves to be a surprisingly competitive candidate for the teen-spirited *@OliviaTaters* too. Surprises are fun, but our metaphor bot will need to explain its reasoning so that the best surprises are not dismissed as the products of random selection. Whichever

generator. The grammar uses transparently meaningful labels for each of its nonterminals, allowing the bulk of its content to be cut and pasted into other bot grammars as you see fit.

A second personality-related Tracery grammar can be found in the *Viral Personalities* directory of our *TraceElements* repository. Bots often reflect the personalities of their creators, so with this grammar we set out to imagine how a computer virus might also bear the mark of its maker, or failing that, the personality of a famous namesake. A key challenge required us to build a mapping of the personal qualities of famous personages to the malicious behaviors of fictional viruses. You will find the products of our wicked imagination in the spreadsheet *viral symptoms.xlsx*. The grammar exploits this mapping as a bridge between the personal qualities of NOC entities to the behaviors of viruses invented by, or inspired by, those entities. The following tweet is typical of the grammar's outputs:

The inspirational virus JE5U5.CHR15T hacks your iTunes account and uses your credit card to buy every Tony Robbins self-help audiobook.

You will find one additional file in this directory: a response grammar that allows CBDQ to respond to Twitter mentions with a tweet like the above if a given property (such as inspirational) is also mentioned.

# 7  Magic Carpets

## The Poetry of the Everyday

In the Coen brothers' 1998 movie *The Big Lebowski*, Jeff "The Dude" Lebowski bemoans the ruination—by urination, no less, and by confused nihilists to boot—of his favorite rug because "it really tied the room together."[1] Our possessions tie us to our past and to each other, and so we often value them more for what they say (or allow us to say) about other things than for what they represent in themselves. The Dude's rug serves a similar role in the film as a whole and is, at one point, used by the Coens as a visual metaphor for the city of Los Angeles.

When casting about for a theme, a poem, a picture, or even just a tweet, we could do worse than to seek inspiration in the detritus of our own lives. As Yeats suggests in his late-period poem about writer's block, "The Circus Animals' Desertion" (where the big beasts represent the noblest and most mythic themes of his earlier works), "Now that my ladder's gone, I must lie down where all the ladders start, In the foul rag and bone shop of the heart."[2] Yeats concedes that his masterful images that "grew in pure mind" owe their genesis to the humblest of sources—"A mound of refuse or the sweepings of a street, / Old kettles, old bottles, and a broken can, / Old iron, old bones, old rags"—and although we can hardly compare the outputs of a bot to the poems of a celebrated poet, we can certainly see similarities in their sources of inspiration. For a Twitterbot, as for Yeats and other human poets, all ladders begin not in a place of pure poetry but in the everyday world of mundane resources and unremarkable possessions. The poet's job, and the bot's programmed goal, is to turn these constellations of objects into occasions of meaning with a text that "really ties the room together."

So what's the Twitter equivalent of Lebowski's rug? We have seen in previous chapters the importance of a good framing device for tying together the various resources that go into a machine-crafted tweet, and the Twitterbot equivalent of a room-defining rug can be as simple as a single hashtag, whether *#JudgeMe*, *#ThingsTrumpNeverSaid* or even *#Fifty-ShadesOfDorianGray*. It can be a syntactic framing that evokes a popular idiom or parodies the folk wisdom of an overused cliché, or it can be anything at all that serves to weave the distinct linguistic and conceptual strands of a tweet into a coherent yarn in the reader's mind. It is a unifying textile, metaphorically speaking, that need not be wholly textual. Save for a brief off-road excursion into the world of emoji in the previous chapter, all of our bots and framing devices have been mostly textual up to now. But Twitter and its API also support the attachment of images to our bots' tweets, allowing, for example, Anthony Prestia's *@greatartbot* to use software built by Andi McClure and Michael Brough to generate new abstract images for its followers.[3] Yet other bots exploit Twitter's multimodality to present readers with meanings that emerge from the yoking together of images and words. Allison Parrish's bot, *@the_ephemerides*, to take just one example, yokes computer-generated poetry (formed by intercutting snippets of a nineteenth-century astrology text with cuttings from an oceanographic text from the same era) to images of our solar system's outer planets as captured by NASA probes.[4] The icy images lend a cosmic grandeur to the texts, while the portentous texts prompt us to look for meanings and resonances in alien images that we might not otherwise perceive. Like any good marriage, each partner brings out the best in the other, or makes the case that we should at least look for the good in the other. When image and text work well in isolation and come together in harmonious potential, the result can be grandly poetic, as in this couplet from Parrish's bot that is tied to a close-up of Jupiter's dark, stormy eye: "Water diminished, where firmly embedded down in / the sea on a dark night, to glow like a white-hot cannon-ball." Parrish's bot embodies the can-do philosophy of bot design by seeking to multiply and magnify the value of preexisting resources, in this case a trove of NASA imagery and a cutup generator fueled by a buddy-cop pairing of delightfully oddball texts. These serendipity-courting heterografts allow one resource to compensate for the other's occasional limits, as when an image that seems dull on its own (a low-resolution closeup of broken lines) is married to this: "They failed in the length at a point to each particle / of an inch; and when highly magnified, to secure."

Parrish's *@the_ephemerides* pairs preexisting images to newly constructed poetic texts, not unlike *@appreciationbot*, which pairs newly minted metaphors to images from a museum catalog, courtesy of yet another thrifty recycler, Darius Kazemi's *@museumbot*. The web is filled with images and texts that are ripe for reuse in new contexts and with new meanings in this way. Nonetheless, in this chapter, we explore a means by which our bots can construct bespoke images of their own, to place into meaningful union with novel texts that—using the techniques of earlier chapters—are also of their own making. We begin where we came in: taking a closer look at the Dude's prized rug.

## Blots on a Landscape

A compelling case for the value of abstract modern art is made by Susie Hodge in her book *Why Your Five-Year-Old Could Not Have Done That.*[5] Yet to be fair, who among us hasn't responded with similar denunciations at some time or other to some piece or other of apparent simplicity, if perhaps in thought rather than word? It does not take much to see how a willful child might slash a canvas in the manner of Fontana or even take pride in this act of wanton destruction. Random pigeon splatter on a felt roof can remind us, if in the right mood, of a Pollock, and a child's rough-edged slabs of color may remind a generous teacher of a Rothko. But if every disrespected surface or discarded urinal were considered art, every dump and dung heap would be a densely stocked museum. Intent, or at least the perception of creative intent, is the essence of art that these random acts of wantonness and waste fail to exhibit. But artistic intent is not a nut we are going to crack here, even if we do attack the problem of novel image generation with the software equivalent of a jackhammer. Our bot's images may resemble those of Hodge's five-year-old more than those of a Riley, Rothko, or Pollock, but as in Allison Parrish's *@the_ephemerides*, our bots can also caption these images with enigmatic texts to create occasions of resonant meaning for us to ponder. As such, we aim to harness randomness to create interesting pictures and hope that a semblance of intent emerges from the apt juxtaposition of words and images.

Let's begin with pure untrammeled randomness and see where that takes us. It is a simple matter to randomly splatter paint onto a pixelated canvas in Java or in any language with basic graphics capabilities. Our program can set the color of individual pixels in a canvas of a given width and height—let's say $1,024 \times 768$ pixels, to give us a reasonably high-resolution landscape—or it can draw and fill overlapping shapes (e.g., rectangles, ovals, triangles) of random size and color and of varying transparency. Whereas the former strategy is likely to give us an indiscriminate fog of pixels, like a TV tuned to a dead channel, the latter often yields Pollock-like images in which stubborn patches of rough color breakthrough.

 **ColdFinger** @BotOnBotAction · 6m
I made this wallpaper from Bureaucrat-grey, Tar-black and Albumen-white. But I now christen it "Stone Chimney Clouds."
#StoneChimneyCloudRGB

This image was composed by naively drawing a great many circles, squares, and triangles of gradually decreasing size, alternating hue and random alpha value (transparency) at random positions on a 1,024 × 768 pixel canvas. (For obvious reasons, we focus our attention here on black, white, and gray examples.) Variable transparency settings are crucial to the painterly effect because we want our stock shapes to overlap in ways that simulate the smooshing of real paint on canvas. Circles mimic blobs dribbled from a brush, and triangles and squares simulate the hard lines of a paint knife. In chapter 5, we discussed how multiple RGB colors can be mixed to produce a single RGB blend, which, in the mold of *@everycolorbot*, might be attached to a tweet as a single homogeneous block of color. Here, in contrast, because the painting process is noisily random, the resulting image is far from homogeneous. Unlike the swatches of pure color produced by those earlier RGB bots, these uneven images allow viewers to easily discern the different colors that contribute to a canvas. By so clearly linking words to images, we allow one modality to inform our appreciation of the other.

Our bot has several freedoms denied to traditional artists applying real paint to real canvas. It can, for instance, wait to decide the color of a pigment until after it hits the canvas. So when splashing paint in a mostly random fashion onto the canvas, our bot can choose its color to suit the area in which the paint lands. Suppose it has two colors at its disposal, red and blue; it can choose red for any splotches that land on the left-hand side of the canvas (or on the top half, or the bottom half, or under the diagonal from top-left to bottom-right, and so on) and choose blue for the splotches landing elsewhere. In this way, even randomly applied paint will result in a coherent image overall, while the random splashes of circular and triangular paint blobs will also create some satisfyingly rough and painterly borders along adjoining areas of color. Here are some examples of the simple splatter-edged patterns that can be made with just two or three colors:

If these images strike you as better suited to a beach towel or a doormat than to a gallery wall, they are interesting enough (for now) to fill the same basic role as Lebowski's rug: paired with the right words, they can tie a tweet together. To start, let's revisit our goal from chapter 5, the creative naming of RGB hex codes, and apply the same naming strategies to these bespoke images. Recall that our dominant strategy was to repurpose well-formed linguistic ready-mades scavenged from a large source of language data, such as the Google web n-grams. Such n-grams must be found a minimum number of times on the web to be considered at all (e.g., forty is the threshold imposed by Google), and their frequencies are themselves a useful source of data regarding a reader's presumed familiarity with any given word combination. Our earlier bots sought out n-grams composed entirely of lexicalized color stereotypes—words that denote ideas or things that are strongly associated with specific colors. We use precisely the same strategy here, though our bot will now use ready-mades to name multihued images rather than monochromatic swatches of an RGB blend. The two images shown here both bear a name gleaned from the Google 2-grams, "piano ivories" (freq. 44) and "ballerina mouse" (freq. 88). The linguistic framing strategy in each case is very much on the nose: the bot states the name it has given to the two-color image, suggesting that a number of followers equal to the n-gram frequency of the 2-gram have supported this choice. Rather than specify the actual RGB codes of the colors involved or redundantly naming the color stereotypes that contribute to the name (e.g., piano black, ballerina pink), the bot casts about in its lexicon for other stereotypes associated with the same hues. In effect, each image is made to bear the weight of multiple visual metaphors. So an image named for piano keys aims to remind readers of dark espresso and bright stars, while an image named for a dancing mouse also becomes redolent of fish.

I composed this wallpaper using Espresso-black on Star-white. Yet when I asked for apt names, 44 of you suggested "Piano Ivories."

I composed this wallpaper using Salmon-pink on Anchovy-brown. When I asked for apt names, 88 of you suggested "The Ballerina Mouse."

The French writer and Dadaist artist André Breton was fond of saying that "art should be as beautiful as the chance meeting on a dissecting-table of a sewing-machine and an umbrella."[6] In the spirit of the ready-made, Breton borrowed this enigmatic line from another writer, Comte de Lautréamont, because it suited his philosophy of contextual dislocation and reintegration so perfectly. Breton saw that familiarity can rob everyday objects of their essential magic, habituating us to seeing them from a single utilitarian perspective. Only by forcing objects into jarring juxtapositions and forcing the viewer to seek out new connections and resonances within these amalgams can mundane objects be made to once again reveal their overlooked potential. Breton's philosophy remains the driving force of a great many Twitterbots, especially those that make a virtue of incongruity by reveling in the seemingly irrational combination of words, images, and ideas. Each of the two-color images above is a dissecting table that is made to carry four familiar (if largely unrelated) ideas apiece, inviting viewers to seek out new and evocative connections that are grounded in their own personal experiences. The chosen name goes a long way toward establishing the bona fides of even the craziest amalgam, but these names, like the ideas they evoke, can themselves be the subject of an exploratory combination process. Consider the following name:

I made this wallpaper from Oxblood-red, Pear-green and Caffeine-brown. But I now christen it "Lollipop Tree Crusts."
#LollipopTreeCrustRGB

Some readers may harbor stubborn memories of "The Lollipop Tree," a children's song made famous by Burl Ives (who also sang "Big Rock Candy Mountain"), and this might explain the bulk of the frequency of this pairing in the Google n-grams.[7] But the idea that lollipop trees might have unpalatable crusts is sourced from another 2-gram, "tree crusts," which shares only its first word with the former's second. So while the 3-gram "lollipop tree crusts" is not to be found in the Google n-grams, or anywhere else at all on the web at the time of writing, it is an invention that is grounded in possibility via these attested, interlinked 2-grams.[8] The three-color image serves as an additional tie on this trio of words and ideas, coaxing readers into finding a coherent unity within this odd, linguistic bricolage.

Ideally, the images that tie a Twitterbot's tweets together should be intricate enough to spur the retrieval of relevant memories and experiences by a viewer and original enough to keep these viewers coming back for more. Our bots may not know what any given image means to its viewers or even understand the semantics of the names that it assigns to its images. Nonetheless, an image of sufficient novelty and intricacy can act as a Rorschach blot onto which viewers can impose their own interpretations, guided in large part by the accompanying text of the tweet. When held to this higher standard, our repertoire of patterns above clearly falls short of the mark. So in the following sections, we explore how our bots might weave complex and original images of their own—images that are not just predetermined blots on a landscape but Rorschach blots on a landscape. We will make our bots responsible not just for weaving these images, but for inventing the rules that give them their internal coherence. Like Lebowski's rug, our bot's images will be woven, line by line, from strands of colored yarn.

## They Live!

The most magical carpets are those that weave themselves, and in similar object-oriented fashion, our images must determine their own patterned use of colors. Our best chance of pulling off this magic trick is to view our self-weaving images as living things, with their own DNA-like codes determining how they unfurl. But how might rules that are rigidly followed like clockwork yield images that are at once unpredictably novel on the outside yet predictably regular on the inside?

Like all living things, this "genetic" code is only part of the story. Even when the code is a deterministic computer program, its rules may be activated in very different environments that push the "organism" in unpredictable directions. The initial conditions that prevail when the code is first run are open to extreme random variation, but here's the thing: the rules may themselves be the product of random selection. To weave a new image, our bot can create a random system of rules and a random starting environment for the rules to operate within. If the image that results is deemed acceptable, it is duly framed by the bot and tweeted. If the image is deemed unacceptable, the bot can simply try again with a new starting environment or a whole new set of random rules, or it may decide to scrub both and start all over again. Our bots thus go from being systems of rules for creating images or texts to systems of rules for creating other systems of rules and the conditions in which to run them. To use this power wisely, a bot must have its own in-built sensibilities, in effect metarules, enabling it to reject rules or conditions that yield substandard images, so that it can single-mindedly search for the pairings of rules and conditions that yield tweet-worthy results.

Who knew clockwork could be so unpredictable? But this is not the clockwork of cuckoo clocks and church towers where cute mechanical figures appear on cue to strike poses, bells, or each other before retreating back into their alcoves. This is the clockwork of mathematician John Conway's "The Game of Life," a grid of autonomous cell-like entities with their own simple but inviolable rules for when to live and when to die.[9] Conway called this arrangement a "cellular automaton." In the simplest arrangement, each cell on a two-dimensional (2-D) grid has eight immediate neighbors, and the state of each cell at time $t + 1$ depends, via rules, on its own state—and the states of its neighbors—at time $t$. The cell at any given position can be in just one of two states, alive or dead, and the rules determine when a cell comes alive and when it ceases to live (though later reincarnation is a possibility). For instance, we can cause cells to die of loneliness with a rule that insists that any living cell at time $t$ becomes dead at time $t + 1$ if it has just one neighbor at $t$ or none at all. A dead cell at time $t$ can come back to life again at time $t + 1$ if it has exactly three living neighbors at time $t$; if this salvation narrative does not satisfy, think of the newly living cell at $t + 1$ as the offspring of its three neighbors. To keep this population of Lazarus-like cells under control, we assert another rule to thwart rampant growth: a living cell at time $t$ becomes dead at $t + 1$ if the unfortunate cell has more than three living neighbors at time $t$. This means that a Goldilocks cell that is alive at time $t$ and has just the right number of living neighbors—no fewer than two and no more than three—will continue to live at time $t + 1$. These four simple rules, applied rigidly at each tick of our system's internal clock to a starting array of cells that are initially either dead or alive, give rise to some truly surprising patterns of cellular activity.

Certain configurations of cells in a world that obeys these rules can be entirely stable, forming what aficionados call still lifes. For instance, a

2 × 2 box of living cells bordered on each side by dead cells will continue to hold its shape, since each of its four members (all corners) has exactly three living neighbors. Other configurations, which are less stable (termed semistable), will regularly oscillate between different patterns of life and death. Groups of cells called "blinkers" oscillate back and forth between orthogonal configurations—say, a vertical and a horizontal bar of cells—on each tick of the clock.[10] However, the most famous oscillators of all are the "gliders," semistable groupings of cells that cycle through a sequence of intermediate patterns before returning to their original configuration. Unlike the simpler blinkers, though, this cycling causes a glider to incrementally shift its position in the grid overall, so that over time, this group of cells appears to glide across the 2-D world of the automaton. The rules of our simple automaton are all local, which is to say that the state of any cell depends on only the current state of itself and that of its adjacent cells. However, gliders create new possibilities for long-distance interaction, since a glider that originates in one area of the grid can make its way to a far distant area and destabilize any still lifes that it meets there. Thus, a glider that approaches our earlier still life of four living cells may puncture its protective wall of dead cells and cause our little hermit kingdom to blossom into life or wither unto death. More intriguing still is the possibility that cells unite to form glider guns, complex clusters from which gliders regularly emerge every thirty or so ticks of the clock.

There is a straightforward mapping from cells to pixels and states to colors: we can map a living cell at position $<x, y>$ to a black pixel at screen location $<x, y>$ and map any dead cell at a specific location to a corresponding white pixel. A 2-D automaton of two states is thus easily rendered as a 2-D image of two colors. Since these automata help us visualize how complex behavior can emerge from the interaction of remarkably few rules that are as simple to design as to run, they are popular as assignments for undergraduates and neophyte coders alike. The web is awash with interactive simulations that permit you to tweak the starting state of different cells or even to define your own simple rule sets. It is no exaggeration to say that a good implementation can draw you in and make you wonder where the day has gone. However, the visual charm of a 2-D cellular automaton resides largely in its dynamism. In a static rendering, a glider gun is merely a cluster of cells, and not a particularly attractive cluster at that. In a dynamic rendering, however, its operation can appear magical as its internal cells organize themselves to launch another glider into the grid before returning to their starting configuration so that the whole cycle of creation can run again and again. So as candidates for the generation of static images that can be tweeted by our bots, these cellular patterns fall

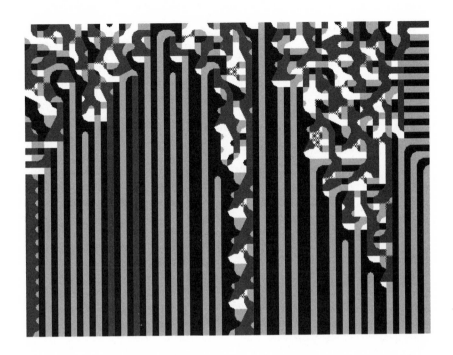

The point remains, however, that even if time-consuming search can sometimes be avoided, critical self-analysis should not. Our bot will still need to determine whether any image it generates—from its first to its 50,000th candidate—meets its aesthetic criteria for a tweet-worthy visual, whatever these might be. In any case, search is easy to perform, and relatively fast too, if a bot initially works at the level of 64 × 48 patterns and limits its appetite to a few hundred thousand candidates. Indeed, brutish search with blind variation typically suffices, with the bot simply generating a new random rule set and a new random row of states whenever its current pairing fails to pass muster. Though such a bot might jump blindly from one area of its vast search space to another, it will soon enough land on an acceptable image if its metarules are not overly demanding. But these bots might also harness randomness in a more intelligent fashion by mimicking the search strategies of predators in the natural world. Because a shark cannot sate its appetite with a single fish, it must apply its energies to the search for an aquatic banquet. It does this by using what mathematicians call a Lévy flight strategy.[12]

Finding itself in a small area of its search space, a vast ocean, the shark will dart hither and thither in search of food, so that its random movements resemble the Brownian motion of a dust mote in an air-conditioned room. After some time, however, if it cannot find sufficient food in its current locality, it does not do anything so dull as to gradually expand the horizon of its search; rather, it boldly strikes out to explore a distant pocket of the space. By interspersing its minor variations with occasional upheavals, sharks instinctively seek out what creativity scholar David Perkins dubs a *Klondike space*.[13] Upon finding itself in a veritable Klondike of high-value targets, our agent—whether a shark, a miner, an artist, or a bot—greedily harvests as many of these targets as it can. Lévy flight is easily interleaved into a bot's existing strategy of blind variation by randomizing the automaton's rules and its starting states on varying schedules. For instance, a bot might randomly generate new start states for every variation, but generate a new rule set only at every $N$th variation (where $N = 100$, say). Our bot would thus consider $N$ rows of random start states for every random rule set it produces.

A more interesting question than where or how a bot finds an acceptable pairing is what it should do when it finds one in a new area of the space. After all, this new area may turn out to be an abundant Klondike of further opportunities. So should our bot prioritize novelty by always seeking out new hunting grounds, or should it linger in this potentially fertile area a while longer to create further images in a similar style? Human artists often choose the latter course by opting to comprehensively explore a distinctive motif or color palette that has yielded a recent success, and there is no reason that a bot cannot likewise have its own "blue period" or its own passing flirtation with bars, stripes, and triangles. Human artists can also be receptive to implicit feedback from their audiences, inasmuch as the perceived success of a work may be as much a matter of public acclaim as personal taste. A bot can also assimilate the preferences of its followers by attentively tracking the kinds of images that are most often "liked" or retweeted.

An obvious application of basic machine learning techniques thus suggests itself: our bot can maintain records as to the frequency with which images with specific features are liked or retweeted by others, and when later searching its vast space of candidate images, it can seek out those that score well on an objective function that prizes the most popular features. These features would likely comprise specific color combinations and generalized automata rules. For instance, a rule matching the schema $aaa \rightarrow b$, where $a$ and $b$ are states and $a \neq b$, generalizes to the motif horizontal stripes; a rule matching $aaa \rightarrow a$ generalizes to expanse; a rule matching $ba? \rightarrow b$, where $?$ can be any state at all, generalizes to right diagonal, and one matching $?ab \rightarrow b$ generalizes to left diagonal; and so on. The coverage provided by a rule for a given image serves as its relative weight for that image, and this will be further weighted by the popularity of the image on social media. With enough user feedback, our bot can learn to produce the images that its followers seem to appreciate the most. Since the active users uncommon to see restrooms labeled "cowboys" and "cowgirls" and "cavemen" and "cavewomen" or to see icons of maleness and femaleness such as top hats and bonnets or bow ties and hair ribbons take the place of the traditional stick figures. But by relying on gender stereotypes, these humorous signs also reinforce our stereotypes about gender. @RestroomGender cleverly avoids stereotyping by eschewing familiar meanings and conventional significations. Instead, the bot's nonsensical labels and abstract visuals can seem like metaphorical provocations that imply much more than the bot actually knows how to say. Like the combinations of five-dollar words in Darius Kazemi's @metaphorminute, its vivid juxtapositions tantalize with their almost-meaningful explorations in the possibilities of language and embolden us to imagine real scenarios in which these possibilities might almost make sense, such as in libraries with special facilities for older members ("Jurassic readers"). Even if the local details are meaningless in themselves, the weaving together of words and images can really tie the restroom together.

Esoteric symbolism and novel images may hold no meaning at all for readers, but they can hold out the promise of so much more if framed with a text that makes readers want to look deeper and see meaning. Our bots stoke this desire to perceive intentionality and meaning by demonstrating a basic fluency with the ideas that matter to human readers, that is, by making their whimsical outputs human scale. Consider how images and text are integrated by @BotOnBotAction, a Twitterbot we met earlier that spends much of its time critiquing the Twitter personality of other bots and the humans who opt in. Occasionally the bot also generates multimodal musings on ideas that might matter to all of us. In the following tweet, it simultaneously renders—in abstract visual terms and concrete linguistic terms—what it might mean for a human to be a "hero." The image itself is a vertically and horizontally mirrored four-state 1-D automaton unfurled over time, in which three of its states correspond to RGB encodings for brown, black, and gray and in which the fourth state is to an equal-parts blend of those three other colors. The resulting abstract image is interesting in itself (though you have to trust us on the colors), but it becomes truly meaningful only in the context of its conceptual brief. Drawing on a database of stereotypical ideas and the qualities we most associate with them, such as heroes, villains, goddesses, bureaucrats, presidents, and dictators, the bot picks a stereotype at random and decides to make its image about that. It does so by taking a selection of the idea's most stereotypical qualities, again at random if there is a surfeit, and goes about finding other stereotypes from which it can build the necessary color palette.

**ColdFinger** @BotOnBotAction · 8m
I call this picture "Hero". I painted it with idealistic Obama-brown, devout priest-black and deadly wolf-grey. #HeroRGB

Consider the three stereotypes evoked in this particular tweet. Obama's inspiring leadership and status as the first African American US president secure him a place not just in the pantheon of leaders but in the bot's color lexicon. Priests have many typical properties, not all of which are positive in light of recent scandals, but devoutness is perhaps the least controversial, and their traditional black uniform gives priests a clear RGB binding in the color lexicon. Finally, since wolves are just as likely to find themselves exploited in our similes for their color (gray) as for their potential to describe human behavior, they also earn an easy entry into the stereotype database and into the color lexicon.

The bot pushes against its confining character limit to allude and evoke rather than explicitly assert. It offers up its picture as a metaphor for heroism by naming it thus, but it does not tell us where this heroism might actually reside in the image. Rather, it motivates us to see a mix of idealism, devotion, and threat by grounding its use of colors in human-scale ideas on which those qualities are also firmly grounded. Brown hues are no more inspiring than gray hues are deadly or black hues are devout, yet the metaphor seeks to establish these bindings by exploiting familiar stereotypes that imply sound perceptual logic at work here. So it establishes its bona fides as a meaningful metaphor but only hints at what its literal

**ColdFinger**
@BotOnBotAction

I painted "Darkness" after reading
@BestOfBotWorlds's latest tweets, using
depressing funeral-black, menacing bat-grey
and gloomy mist-grey.

Handles can be deceptive, and much of the fun in analyses like these comes from the mismatch between perception and preconception. *@BestOf-BotWorlds* really is a gloomy guts for whom the name is at best aspirational and at worst ironic. Likewise, filmmaker David Lynch may be loved for his darkly twisted worldview, as evident in *Lost Highway* and *Blue Velvet*, yet *AnalyzeWords.com* often scores *@David_Lynch*'s tweets in the high 90s for upbeatness. His tweets are thus more likely to suggest *Sunflower*-yellow, *bunny*-pink, and *Gumby*-green to *@BotOnBotAction* than *Tarantino*-gore or *Sith*-black. Lynch uses Twitter to nurture new talent and promote new projects, and so his account is situated very far away from his twin peaks of death and darkness, Of course, a bot that dishes out the metaphors must also be able to take them, so this is the self-portrait that emerges when *@BotOnBotAction* sings *#PaintMySoul*:

 **ColdFinger**
@BotOnBotAction

⚙ **Following**

I painted "Outcast" after reading @BotOnBotAction's latest tweets, using isolated prison-orange, lonely wolf-grey and reclusive yeti-white.

   @*BotOnBotAction*'s strategy of not addressing other users directly, even when they opt in with the hashtag #*PaintMySoul*, means that the bot does a good deal of social name-dropping yet scores very poorly for sociability. So its own view of itself as a reclusive outcast, hurling painted rocks at happy townspeople from its remote perch, is far from inaccurate. But what if the bot were to enlist the services of human color experts on social media, who, as we have seen, clamor to offer color recommendations on *ColourLovers.com*? In addition to allowing users to name and express "love" for RGB codes, the site also encourages color lovers to assemble and give meaningful names to palettes of their own design. We have downloaded twenty thousand of these named palettes—with a name and a set of five RGB codes apiece—to use them as a source of guidance to the bot (the list is on our GitHub). Simply, the bot uses a given user palette to paint its latest image and applies the name of the user's palette to the resulting work. As though working for a patron, the bot then dedicates the work to its human muse:

Vogler's twelve steps seem too linear or too coarse, an alternate system proposed by the Russian folklorist Vladimir Propp in his 1928 work, *Morphology of The Folktale*, offers a freer and more granular picture of the relationship between character and plot.[15] Folklorists are empiricists at heart, and Propp built his system of recurring story elements, or story *functions*, from a painstaking analysis of a corpus of Russian tales. In contrast to Vogler's twelve steps, Propp identified thirty-one recurring elements in his analysis, which he arrayed into an idealized sequence that is far from rigid. The earliest functions in this sequence loosely align with the earliest of Vogler's steps, and we can see how Propp's functions Abstention (a key member of the community suddenly leaves, perhaps unwillingly), Interdiction (an edict or prohibition is placed upon the community, curtailing its freedoms), and Violation (an edict is violated, incurring the wrath of its issuer) might motivate a hero to heed Vogler's *call to action*. But Propp also allows the villain to enter the fray during this opening act, via a range of character functions that hint at future wickedness; these include Reconnaissance (the villain seeks out a MacGuffin and forms a plan that will affect the hero and/or the community), Trickery (the villain obtains important leverage by deceiving a dupe), and Delivery (the villain obtains that all-important MacGuffin that will drive the plot forward). The plot thickens when the hero crosses the threshold into adventure via the Departure function or the villain crosses his own threshold of wickedness to impose a Lacking condition on the hero's world, by, for example, abducting a loved one, stealing an object of value, or foisting famine or discord or slavery on the community. If *The Matrix* seems to be shaped with Vogler's twelve-step cookie cutter, it is Propp's thirty-one functions that give the film its specific fillings, as each of its major characters—Neo, Morpheus. Trinity, Agent Smith, the Oracle, and the traitorous Cypher—fulfills a different functional need as identified by Propp in his 1928 study of folktales.

Other folklorists have doubled down on Propp's approach, to bring ever more zeal to bear on the deconstruction of myths and folktales from diverse cultures. The work of folklorist Stith Thompson in the 1950s at the University of Indiana is especially notable for the scale of its analysis.[16] Thompson and his colleagues set out to build a comprehensive catalog of the motifs that recur throughout the world of myth and fable. Their catalog is hierarchical and organizes its motifs into families of generic schemas and specific instances, assigning a Dewey Decimal–like code to each. You can browse the fruits of their labors at the multilingual folk tale database (MFTD) at *mftd.org*. The catalog's contents make for an engrossing read of the "you couldn't make this stuff up" variety, for when shorn of their narrative contexts, the motifs at the heart of so many fables can seem so alien that—dare we say it—they might even be *machine* generated. Consider a motif that the MFTD labels *B548.2.2.2: Duck recovers lost key from sea*. This is cataloged as a special case of *B548.2, aquatic animal recovers object from sea,* which is, in turn, an instance of *B548, animal recovers lost object,* and of *B54x, animal performs helpful action.* A resource as comprehensive as the MFTD allows folklorists to precisely codify the points of overlap between the tales of different cultures, but it can also be used to stimulate the generation of new stories or perhaps suggest motifs and writing exercises for the sufferers of writer's block. In much the same spirit as Darius Kazemi's *@museumbot*, which tweets random samplings from the Met's art catalog, a bot named *@MythologyBot* (courtesy of *@BooDooPerson*) tweets a random pick from Thompson's index of folk motifs at three-hourly intervals. Leveraging the weirdness of the MFTD, the bot dares its readers to dismiss its tweets as machine-crafted cut-ups of more sensible texts. It offers frequent and vivid demonstrations of a counterintuitive truth: we can use precooked schematic forms to tell stories that are traditional *and* oddly original.

It's no accident of language that reporters speak of newsworthy events as "stories." Reporters should not invent the facts, but we do ask them to interpret what facts there are and spin these into a coherent and compelling narrative. Reporters adhere to their own storytelling principles, such as "don't bury the lead," yet they also share many of the same concerns as a writer of fiction. Fact shapes fiction, but the inverse is also true: the revolving door between art and reality ensures that one is always a constant source of inspiration for the other. Newsmen learn from novelists, and storytellers take inspiration from the news. Dick Wolf, the creator of so many TV shows with plots ripped from the headlines (such as the various long-running *Law and Order* franchises), has spent decades mining the news for gripping drama, but he is not the first to do so, nor is he the first to construct a pipeline between news and drama on so commercial a scale.[17] During the 1920s, while Vladimir Propp was conducting his scholarly analysis of Russian folktales to see what made them work as stories, a Canadian writer of pulp fiction named William Wallace Cook was developing a way of synthesizing new plots for his books, which he wrote at speed in a triumph of quantity over quality. Cook's goal was to systematize the process of pure plot creation so that writer's block would never prevent him from meeting a deadline again. He called his system Plotto and championed it as a means of plot suggestion with which writers could quickly generate high-level plot skeletons for their stories.[18]

Every story needs a conflict—note how Propp and Campbell/Vogler are as one on this issue in their analyses—and so the Plotto system sees stories emerge from the combination of themes (or what Cook called *master plots*) and conflicts. As Cook put it in his 1928 book for budding "Plottoists" (his term), "Each master plot consists of three clauses: An initial clause defining the protagonist in general terms, a middle clause initiating and carrying on the action, and a final clause carrying on and terminating the action." We might see that first clause as cueing up Vogler's first four steps (the call to adventure), the middle clause priming the middle stretch of Vogler's steps (crossing into a world of adventure and ordeal), and the final cause as encapsulating Vogler's final four steps (hero's reward and the journey home). But unlike Vogler, Campbell, and Propp, Cook saw Plotto as a practical resource for budding writers, a trove of master plots that he himself had assiduously scribbled in notebooks, clipped from newspapers, cribbed from history books, or distilled from the work of others. His book enumerates more than one thousand master plots, some stale and stodgy and some that look as alien as Thompson's folk motifs when formulated in Cook's concise yet florid prose. And it does not end there: Cook corrals his master plots into a comprehensive system of cross-indexing that allows plot elements to be colored by different conflicts and clicked together like LEGO blocks. Consider the master plot numbered 1399:

> A seeks wealth, his by right, which has been concealed * A seeks wealth which his father, F-A, has left him, but concealed in a place whose location has been lost

Cook uses placeholder variables *A* and *B* to denote, respectively, male and female protagonists, while placeholders like *F-A* above denote character functions such as "*A*'s father." The plot has both a generic and a more specific rendering, separated with the * token. Cook indexes his master plots by conflict type, and he places the plot above in group 57, "Seeking to unravel a puzzling complication." He cross-indexes each plot to others so that writers can connect their plots like the track segments of a train set. Cook links master plot 1399 to this potential follow-on segment:

> A asks that B allow herself to be hypnotized in order that he may learn where buried treasure has been concealed * A hypnotizes B, and B dies of psychic shock

This is in turn linked to the following master plot, suggesting a dark closing act:

> A helps A-2 secure treasure in a secret place * A, helping A-2 secure treasure in a secret place, is abandoned to die in a pit by A-2 who makes off with the treasure

The wonder of Plotto is not its plots per se, which can read to the modern eye like the stuff of Victorian bodice rippers, but Cook's system of plot organization. Just as Flann O'Brien invented a proto-hypertext with *At Swim-Two-Birds*, Cook's Plotto is very much a steampunk imagining of symbolic AI in the 1960s and 1970s, and it is, in its way, an application of what Ada Lovelace called "poetical science." Like O'Brien's tongue-in-cheek views on the construction of intertextual collages with precooked characters, Cook's proto-AI also offers an early vision of the cut-up method of text generation that Brion Gysin and William S. Burroughs would later make famous, though Cook's version is much more tightly constrained and bureaucratic in spirit.[19] Yet there is also something of the Twitterbot spirit in Cook's Plottoist approach to the synthesis of novel human experiences via mechanical methods. Cook's own stories may not have stood the test of time, but with access to tools like Tracery and Cheap Bots Done Quick, he might have built some remarkable Twitterbots.

### The Hero and Villain with 800 Faces … and Counting

For Joseph Campbell, the mythic hero figure is a recurring archetype that pops up in countless guises in just as many tales of popular mythology. Whether we pick Gilgamesh, Rama, Beowulf or Conan, or Samson, Moses, Joan of Arc, or Jesus, or Allan Quatermain, Indiana Jones, or Lara Croft, or Sam Spade, Philip Marlowe, Jane Marple, Lemy Caution, Rick Deckard, or Jeff Lebowski, these characters all have the right stuff to undertake a heroic journey for us and with us. These, and many more besides, all reside in Flann O'Brien's archetypal limbo (population: untold thousands) "from which discerning authors [can] draw their characters as required." One digital realization of O'Brien's limbo is Wikipedia/*dbpedia.org*,[20] or even *TVTropes.org*,[21] but another more amenable version is the NOC list, which gives our bots access to as many heroes or villains or sidekicks or mentors or false friends or love interests as a story-generation system could hope for.

Recall that the NOC list offers up positive and negative talking points for each of its more than eight hundred residents, so that each has the background to play a flawed hero or a redeemable villain. The qualities that establish a character's heroic standing, as well as those that establish a character's lacking, to use a term from Propp, are all there waiting to be corralled into a brand-new story. The NOC list is also a well-stocked props department that provides all the necessary costumes and other accoutrements to our automated raconteurs so they can establish a vivid mise-en-scène for a story. Let us assume, for simplicity, that each story will be woven around of pair of two NOC characters *A* and *B* (unlike in Plotto, the letters do not imply a gender). The first of these can be plucked at random from the NOC list, but the second should be chosen so as to exhibit intentionality and create the conditions for an interesting story. However *A* and *B* are chosen, they comprise the flip sides of a narrative coin that will spin continually as the tale is told, with each turn highlighting the qualities and actions of an alternating face. Our stories can reimagine the past or the present by choosing *A* and *B* to be characters that are already linked in the NOC list. If linked via the Marital Status dimension, *A* and *B* will be characters that are known to have married, divorced, or just dated, and if via the Opponent dimension, *A* and *B* will be characters that are known to be rivals. These are first-order connections, insofar as the link between *A* and *B* is asserted explicitly within the knowledge base. Thus, a bot might weave a story about Cleopatra and Julius Caesar, or Angelina Jolie and Brad Pitt, or Lois Lane and Superman, yet do so in a way that completely reimagines their relationships (because, to be frank, the bot will not know enough to faithfully reimagine them as we all know them). Like the first row of a cellular automaton, the relationship between *A* and *B* establishes the foundation on which the ensuing story will rest, so this pairing should be selected with care. Consider the following setup from a storytelling bot *@BestOfBotWorlds*, which sets out to reimagine an old rivalry:

**For Fun and Prophet**
@BestOfBotWorlds
        ☼  👤+ Follow

What if a , who thought it was Thomas Edison, was sued by a  who thought it was Nikola Tesla?

Let's put to one side for now the question of why emoji animals are used here for the famous rivals Edison and Tesla, noting only that emoji are useful single-character icons for complex *A*s or *B*s and that when the relationship at the heart of a story is a factual one, as it is here, framing it as a "what-if" scenario via emoji ensures the counterfactuality of the narrative. This is a signal to readers that the story is a playful reimagining of history with only a fabulist's regard for the truth.

Just as Godard picked Tarzan to oppose IBM, a bot may choose its $A$ and $B$ to serve as vivid incarnations of two opposing qualities, so that some quality of $A$ highlights the opposing quality in $B$. With this strategy, a bot might pair the dirt-poor Bob Cratchit with any of the fabulously wealthy Lex Luthor, Warren Buffett, Bruce Wayne, or Donald Trump. Or a bot might match the well-mannered Emily Dickinson with the more vulgar Eminem, or pit the savage Conan the Barbarian (or Tarzan, for that matter) against the urbane and sophisticated Gore Vidal. The postmodern humor of Godard's pitting of the fictional Tarzan against the very real IBM can also be facilitated using the Fictive Status dimension of the NOC list, so that, for example, the spiritual and kindly Mahatma Gandhi is pitted against the (darkly) spiritual and malevolent Darth Vader. A pairing based on inferred opposition is a second-order connection between characters, insofar as the link is not directly provided by the knowledge base and must be discovered by the bot itself, in a search of the NOC list's unstated possibilities. Naturally, given the combinatorial possibilities that a search can consider, the space of second-order connections is far larger than that of first-order connections, and a bot can make greater claims to originality by exploring a large second-order space of its own construction than a small first-order space that is given to it on a platter by its designer. A host of second-order spaces can be mined to obtain resonant character pairings; some spaces are simple and based on an obvious premise, and others can be far more complex. Consider the space of character pairs that just share NOC talking points. These pairings are essentially metaphors, but metaphors that suggest apt and imaginative story possibilities. So consider another pairing for Nikola Tesla:

**For Fun and Prophet**
@BestOfBotWorlds

⚙

What if a reclusive , who thought it was Nikola Tesla, hid from a 🕊 who thought it was Doc Emmett Brown?

The nutty Doc Brown of the *Back to the Future* movies seems an ideal fictional counterpart for the real life Tesla, for Tesla was something of a nutty professor himself.[22] It is often said that heroes are only as great as their opponents allow them to be, and in the hero's journey, we require

batch. This pairing arises from an obvious second-order space that links famous people portrayed by the same actor, but the actions used to link these characters and drive the plot forward must come from a specific understanding of the characters themselves. Clearly, Dr. Strange is a doctor in the NOC list and diagnosing others is just what doctors do, so this pairing would aptly fit the motif "doctor diagnoses patient" if only we had a stock of motifs like this for our bot to exploit.

But just as Stith Thompson and his colleagues had to knuckle down and build their database of motifs from scratch, we too shall have to build this inventory for ourselves. Our job is a good deal easier, though, because we are inventing rather than analyzing and we do not have to trawl through the world's collected folklore. As motifs are schematic structures that concern themselves with character types rather than character specifics, our first order of business is to create an inventory of the pairings of character types that will be linked by these motifs; then we can set about the task of providing specific linking verbs for the types paired in each motif. To construct this inventory, we consider the pairings across all of the first- and second-order spaces we plan to use for our stories and generalize each to the type level using the NOC *Category* dimension. For instance, we find Forrest Gump + Robert Langdon in the space of people linked by a shared actor, and one way that this generalizes at the type level is *Fool + Professor*. Once this inventory of paired types is created, we can sort it in descending order of coverage, so that the motifs that cover the most character pairs are pushed to the top. We then work our way down the list, providing linking verbs for the character types in each generic motif; for *Fool + Professor* (if we make it that far down the list), we can provide the verbs "study under" or "look up to" or "disappoint." Readers can find a version of our motif inventory with linking verbs for more than two thousand type pairings on our GitHub, in the spreadsheet named *Inter-Category Relationships*. Take this as you find it, or adapt it to reflect your own intuitions about narrative.

### The Road to Know-Where

Our characters are paired on the assumption that similarity is most interesting when there are so many reasons *not* to take it seriously. Doc Brown is a comedic fiction, but Nikola Tesla was a very real and tragic figure, and though Leonardo da Vinci and Steve Jobs were each the real deal, they lived in different historical eras. Emma Bovary and Alice in Wonderland are both wholly fictional beings, yet the fact that each was portrayed by the same actor undermines the credibility of the conceit as a serious story idea.[23] Each pairing is as much a conceptual pun as a conceptual metaphor, but that's also the source of its appeal: these star-crossed pairings are designed to tickle the fancy of a bot's followers, not to pitch woo at a movie studio that might turn them into expensive cinematic products. Yet this is not to say that our bots shouldn't take their own story ideas seriously. After all, a story idea is only as good as the stories that can be woven from it. An inventory of schematic motifs gave our story bot its initial pitch for a story in a single tweet, by providing—in an apropos plot verb—a vivid sense of how its characters might interact. But now our bot must build on these premises to generate full stories that can stretch across many threaded tweets. Let's begin with the question of where a bot will find plots to sustain its stories. The answer is an oldie but a goodie: we're going to treat every story as a journey.

It is not just heroic quests and road movies that build stories around journeys. The language of narrative encourages us to speak of all stories as journeys. So we talk of fast linear stories and slow, meandering ones; stories that take us on an emotional roller-coaster ride; stories that go nowhere, or stories that lose the plot and get stuck in the weeds; stories that race along at a breakneck pace, or stories that just seem to crawl by; stories filled with sudden twists and unexpected turns, as well as stories that lose all momentum before limping across the finish line. As variants of the journey metaphor go, Stories Are Races is especially productive. Actors speak of their most promising projects as "vehicles," and successful well-crafted vehicles do seem to run on fast tracks and turn on greased rails. To lend this race metaphor a literal reality, think of the electric slot car sets that kids have played with for decades. Even if you haven't played with a set yourself, you will almost certainly know of kids who have. Dinky little toy cars are slotted into current-carrying grooves in a track made of pre-fabricated segments, allowing the electricity-powered cars to zip around the track in a thrilling simulation of a real high-speed car race. The goal is to beat your opponent in the parallel groove, and the trick is to modulate your speed so that your car doesn't fly off the track on a tight bend or at a chicane. The more complicated the shape of the track, the more dramatic the miniature narratives that a child can concoct. But complex track con-figurations need a great many prefabricated shapes to click together to form a circuit, and to support certain kinds of dramatic possibility, a child will also need specific kinds of track segment. For instance, two cars will always run in parallel grooves, no matter how many twists or turns in the track, if the track lacks a piece in which its two grooves cross over. With this piece, two cars might actually crash into each other as they switch lanes. But without this, no crashes!

The most popular brand of slot car sets in Europe is Scalextric, while the Gaelic word for story is *scéal* (imagine Sean Connery saying "scale"), so our bot-friendly implementation of the Stories Are Journeys metaphor and its variant, Stories are Races, has been christened Scéalextric.[24] Beneath the cute name lies a surprisingly systematic analogy between storytelling and racing simulations. Take the two characters *A* and *B*: let's keep the simplifying assumption that each story is built around a well-matched pairing of a protagonist *A* to an antagonist *B*, and so we can take *A* and *B* to be the story equivalent of two cars racing along parallel grooves on the same track. This track is the plot, a sequence of actions that each character must pass through in the right order, as each action frames an event in which both characters participate together. If *A* is selling, then *B* is buying, and if *B* performs surgery, it is because *A* is unwell. Because the same event can be viewed from the perspective of different characters (for example, a *lend* event for *A* is a *borrow* event for *B*), a well-crafted plot arranges its actions so as to draw the reader's attention back and forth between characters, as though the reader was watching a fluid game of tennis. When rendered into English and threaded into tweets, the sequence of plot actions will describe how *A* and *B* proceed neck-and-neck from the starting position of the first plot action to the story's finishing line.

Though our plot will be built from individual actions, like the click-and-extend segments of a Scalextric track, we will group these actions into standardized triples with a uniform three actions apiece. We can think of the action triples that schematize an unsurprising plot development (*X* happens and then *Y* happens, to no one's surprise) as linear track segments, and those that suggest a surprising turn of events (*X* happens but then *Y* happens, defying expectations) as curved segments. An interesting plot, like an interesting racetrack, balances the straight with the curved in a satisfying whole that is neither too predictable nor too zany. To build a plot as one builds a model racetrack, a storyteller must choose compatible action triples to click together like so many Scalextric track pieces. The criteria for well-formed triple combination are twofold and simple: the third and final action of the first triple must be identical to the first action of the triple that will succeed it in the plot; except for this point of overlap, there can be no other action that is shared by both triples (in this way plot loops are avoided). Consider this standard action triple, which might link Drs. Strange and Turing:

| *A* diagnose *B* | *B* trust *A* | *A* operate_on *B* |
|---|---|---|

For convenience we will assume the presence of *A* and *B* in our triples, so this is:

| (1) | diagnose | *trust | operate_on |
|-----|----------|--------|------------|

Event fillers are always assumed to be *A* (subject) and *B* (object) unless the verb is marked *, indicating a reversal of roles (making *B* the subject and *A* the object). Consider another pair of triples that reflect recurring plot structures in stories:

| (2) | examine | *confide_in | diagnose |
|-----|---------|-------------|----------|
| (3) | operate_on | *believe_in | cure |

We can combine triples 1 and 3 in that order because the last action of 1 is the first action of 3 and the triples share no other overlaps. We can likewise combine 2 and 1 in that order, but we cannot combine 2 and 3. When two triples with an overlapping action are combined in this way, what results is a sequence of five successive actions (as we count the shared actions only once):

| (2)+(1) | examine | *confide_in | diagnose | *trust | operate_on |
|---------|---------|-------------|----------|--------|------------|
| (1)+(3) | diagnose | *trust | operate_on | *believe_in | cure |

Triples can be connected into ever-larger chains of story actions, as in:

| (2)+(1)+(3) | examine *confide_in diagnose *trust operate_on *believe_in cure |
|-------------|------------------------------------------------------------------|

Two connected triples yield a sequence of five actions. If we join three triples together, a sequence of seven actions is obtained, and if we connect four, a story of nine actions emerges. So our storyteller need only add more triples to the plot until its desired story length is achieved. But notice how actions are not allowed to reoccur anywhere in the resulting chain. Though actions do repeat in real life, context can make them mean different things. Because context is an issue that is hard for a bot of little brain to grasp, it is best to prohibit recurrence altogether, to avoid the formation of troubling *Groundhog Day* loops. Note also how our triples have been crafted so that alternating actions tend to switch a reader's focus from *A* to *B* and back again. Because triples are linked by a shared action and framed by a shared viewpoint at the point of connection, it follows that when viewpoint alternation is obeyed within action triples, with each triple passing the baton of its focus to the next, alternation will also be enforced at the overall plot level. The Scéalextric triple-store can be found in the GitHub resource *Script Mid-Points.xlsx* in a simple three-column tripartite structure. Simply, each triple is assumed to have a midpoint action, a lead-in action before this point and a follow-on action after this point. As an illustration, here is a peek at the first few rows of the resource:

| Before Midpoint | Midpoint | After Midpoint |
|---|---|---|
| mock | are_forgiven_by | admire |
| are_trusted_by, are_intrigued_by, are_welcomed_by, | spy_on, travel_with, accompany | are_shocked_by |
| tend, care_for | are_pushed_too_far_by, are_disgusted_by, are_shocked_by | abandon, testify_against |
| capture | are_bitten_by | abuse, kill |
| spy_on | are_discovered_by | beg_forgiveness_from |
| exploit | are_feared_by, are_distrusted_by | alienate |
| supervise, look_after | are_trusted_by, are_respected_by | indoctrinate |

Each row stores one or more triples, with disjunctive choices for the before, midpoint, and after actions separated by commas within cells. The third row in the table thus defines twelve unique triples that chart *A*'s movement from carer to skeptic to enemy. Why *A* cares in the first place, or how *B* will respond to *A*'s betrayal, are parts of the story that we must look to other triples to flesh out.

All of these plot triples can be collectively viewed as a graph, a dense forest of branching pathways in which any triple $\alpha$:$\beta$:$\chi$ is tacitly linked to every other triple starting at $\chi$ or ending at $\alpha$. So a random walk into this woods can give us our plot $\alpha$, ... , $\Omega$: starting at a triple containing the initiating action $\alpha$ (as dictated by our choice of characters), a system picks its way from triple to triple to finish at some as-yet-undecided triple whose third and final action is $\Omega$. The chain of actions that leads the teller from $\alpha$ to $\Omega$ then provides the plot skeleton on which a story can be fleshed out. A random walk from any point $\alpha$ may take a wanderer in this forest to many different $\Omega$'s by many different routes, provided it takes care to avoid loops and obeys the basic rules of nonrecurrence. But how dense with pathways should this forest of branching possibilities be so that every walk in the woods is assured of charting a different story? Our forest must allow for many thousands of possible pathways between diverse points of ingress and egress. However, the number of possible triples in the forest is limited by the size of our inventory of plot verbs from which each triple's tripartite structure is filled. Moreover, only a small fraction of possible triples are actually meaningful in any causal sense or show any promise as a story fragment. We shall thus need a relatively large stock of action verbs to compensate for the selectivity with which they are combined into triples. The Scéalextric core distribution (which can be found on our GitHub; see *BestOfBotWorlds.com* for a link and more detail) comes ready-stocked with about eight hundred plot verbs and approximately three thousand plot triples that pack three of these verbs apiece. With more than eight hundred verbs to choose from, the action inventory does not lack in nuance, and many entries are near, but not true, synonyms of others; for example, triples can employ "kill," "murder," "execute," or "assassinate" to suit the context of a story (e.g., who is doing the killing and who is being killed, or why?). A good many verbs are also present in passive forms, allowing a plot to focus on either the agent or the patient of an action.

This nuance proves to be of some importance when we consider the rendering of stories at the narrative level. A computer-generated story is more than a list of verbs placed into causal sequence. We set aside for now the tricky question as to whether any arbitrary pathway $\alpha, \dots, \Omega$ can be considered a "story," and whether any $\alpha$ at all is a viable starting point, or whether any $\Omega$ is a viable finishing point. Rather, let's assume that any path $\alpha, \dots, \Omega$ can be rendered at the narrative level to become a story. Rendering is the process whereby stories go from logical skeletons to fully fleshed out narratives, where terse possibilities such as *A kill B* take on an expanded idiomatic form. Each action is trivially its own rendering, as we continue the long tradition in AI research of choosing our logical symbols from the stock of English words. Thus, *A kill B* might be rendered directly as *<A> killed <B>*, where *<A>* and *<B>* are placeholders where the eventual characters such as Tony Stark and Elon Musk will be inserted to yield "Tony killed Elon." But this staccato does not remain charmingly Hemingwayesque for very long. It's better to choose from a range of idiomatic forms when rendering any given action, since part of the joy of storytelling (and story *hearing*) is the use of words to convey attitude in a teller and stir feeling in an audience. Consider again the plot verb "kill." This might be rendered idiomatically in any of the following ways:

*A kill B* → *A* stabbed *B*, *A* mauled *B*, *A* put poison in *B*'s cup, *A* put poison in *B*'s food, *A* savaged *B*, *A* put *B* in the hospital, *A* gave *B* a terrible beating, *A* punched and kicked *B*, *A* gave *B* an almighty wallop, *A* kicked *B* into next Tuesday, *A* stomped all over *B*, *A* gave *B* a good kicking, *A* viciously assaulted *B*, *A* launched an assassination attempt on *B*, *A* wanted to kill *B*, *A* choked the air out of *B*, *A* flayed *B* alive, *A* knocked the stuffing out of *B*

In a story of just two people, the action "kill" has especially severe consequences, as it is most likely going to rob our narrative of one of its principals. This is not the kind of action we expect to see at the start or even in the middle of a narrative, as our story must still go on with both characters even if one of them is now dead. The renderings for "kill" above alleviate this burden by treating the plot verb as either an expression of homicidal intent (*A* wanted to kill *B*) or as a hyperbolic turn of phrase for an act of grievous rage. If *B* remains alive in the next action, the reader will know that *A*'s deadly intent has not been realized, but if *B* is obviously deceased in the next action (say because *B* now haunts *A*), readers will infer that *A*'s actions were fatal to *B*. Rendering buffs and varnishes a plot structure to give it both nuance and dramatic effect, but as we can see for "kill," it may also use understatement, euphemism, and deliberate ambiguity to diminish the brittle certainty of a dry logical form. Our bots work best when, like good storytellers, they suggest more than they actually say and allow the reader's imagination to do most of the heavy lifting. The rendering of individual actions in isolation, one action at a time, is a simple context-free approach to a problem that is inherently context sensitive, so it is important that our renderings are stretchy enough to fill any gaps that are left exposed between plot actions.

For each plot verb in our inventory we must provide a mapping from its logical form (e.g., *A* kill *B*) to the kind of idiomatic phrasings provided for "kill" above. As with many other things in language, the frequency with which different plot verbs are used in our triples follows a power law distribution, with a small number of popular verbs (such as "trust") appearing in a great many triples and a longer tail of many more (such as "ensure") appearing in very few. This is in part a function of the verbs themselves and their portability across domains, and it is in part a reflection of the mind-set of the triple-store's designers, for it surely says something about us as creators of Scéalextric that its most frequently occurring verbs are "trust," "disrespect," "condescend to," "deceive," "disappoint," "fall in love with," "fear," "impress," and "push too far." Yet whatever verbs turn out to be most useful, we must aim to provide the most renderings for the most popular verbs. Those will recur time and again in our stories, but diverse rendering can introduce variety at the narrative level and soothe the reader's sense of tedious repetition. The GitHub resource *Idiomatic Renderings.xlsx* contains all of Scéalextric's mappings from all of its eight hundred plot verbs to colorful colloquialisms, providing more mappings for the most popular verbs (such as "trust," "deceive," and "disappoint") to allow greater variability in rendering across stories with common actions. If the storyteller chooses randomly from the action renderings available to it, it can ensure that readers are not bombarded with the same clunky boilerplate in story after story.

Rendering shapes how readers will perceive, process and appreciate a plot, and even the simplest one-to-many mapping from plot verbs to idiomatic templates can introduce vividness, personality, and drama into a narrative. Consider the use of dialogue: the old storytelling maxim is, "show, don't tell," so why tell readers than *A* insulted *B* or that *B* complimented *A* when we can show *A* actually saying something offensive to *B*, or show *B* liberally applying butter to *A*'s ego? Later we'll explore how a bot might invent its own generous compliments and scornful insults as they are needed, to leverage what it and its readers already know about their target. For now, we can simply build the dialogue into the mapping of verbs to idiomatic templates. Consider Scéalextric's mappings for "disappoint":

*A* **disappoint** *B* → *A* thoroughly disappointed *B*, *B* was very disappointed in *A*, *B* considered *A* to be a big disappointment, *B* thought "What a loser" when looking at *A*, "Could you be a bigger disappointment?" asked *B* sarcastically, "I'm very disappointed" said *B* to *A*, "I've let you down" apologized *A* to *B*, "You've let me down" said *B* plaintively, *B* considered *A* a loser, *B* treated *A* as a failure, *A*'s flaws became all too apparent to *B*, *B* wrote *A* off as a loser, "What a LOSER!" said *B* to *A* dismissively

This writing is a very long way from Jane Austen, but even occasional snatches of canned dialogue can help to draw readers into a story and make the plot feel that it is unfolding in real time. Of course, the most vivid way of showing and not telling is to use pictures instead of words. A storytelling bot might attach images to its tweets to illustrate the corresponding plot actions, but which images? A convenient source of storybook illustration can be leveraged from emoji, as those simple images have a suitably cartoonish aesthetic for our Twitter stories and can be inserted directly into a tweet, or, for that matter, into the idiomatic mapping of a plot verb. In fact, it is possible to construct an idiomatic mapping entirely from emoji,

**For Fun and Prophet**                          ✿
@BestOfBotWorlds

## So at first, Steve the  funded Leonardo the ✌'s forays into pioneering new technologies

With the opening bookend in place—it's all a marriage of convenience based on money—our teller has laid the foundation for the first act proper of its narrative. But notice how the rendering above seems so oddly apt for the role of Steve Jobs. Rather than rely on the stock idiomatic renderings of its lookup table, the teller has used specific information available to it (from the NOC list) about the character of Jobs, namely, that one of his Typical Activities is "pioneering new technologies." In this way the storyteller contributes to the mise-en-scène of the piece, much as the rainy nights, neon signage, and seedy locations establish the mise-en-scène of *Alphaville* and *Blade Runner*. Our lookup table of idiomatic phrases is used not as the foreground of the rendering process but as a backstop when action-specific rendering fails to produce a text that integrates specific details from the NOC. So the next tweet in the story has Leonardo lie to Jobs, but this not the idiomatic rendering of the action "deceive." Rather, it alludes directly to Leonardo's goal:

**For Fun and Prophet**                          ✿   ⚬ Follow
@BestOfBotWorlds

## But Leonardo the brilliant ✌ deceived Steve the 🍎 about his true intentions when pioneering new technologies

Notice also that Leonardo is qualified here as "brilliant" in the context of his role as a deceiver. The resource *Quality Inventory.xlsx* provides a mapping from plot actions to the specific qualities of the agents and patients that facilitate them. The action "deceive" is facilitated by the qualities "two-faced," "insincere," and "dishonest" in the deceiver, while a patient is more easily deceived if the agent is "brilliant." No one likes to be deceived, and the bot's triple-store suggests Jobs's natural reaction (notice that an emoji translation is included only if there is space in the tweet):

**For Fun and Prophet**
@BestOfBotWorlds
⚙  **Follow**

So Steve the demanding 🐬 left Leonardo the 🎭 hanging for what was owed it.

This is a stock rendering, straight from the bot's lookup table of plot verbs to linguistic templates. But the action that follows this is very specific to Leonardo:

**For Fun and Prophet**
@BestOfBotWorlds
⚙  **Follow**

Well, Leonardo the 🎭 threatened Steve the 🐬 with talk of shredding him with his scythed chariot

Any plot verb that might incorporate elements of an agent or patient's entries in the NOC list is amenable to this kind of specialized rendering. Violent actions are the obvious go-to here, as these can directly avail of the Weapon of Choice field. But other kinds of action can exploit other fields too: actions involving travel and avoidance can avail of the Vehicle of Choice field and the various Address fields, while creative use can also be made of the Creator, Typical Activities, Group Affiliation, Seen Wearing and Marital Status fields in the right action contexts. Here the address field is used to add local color to the plot action *hide_from*:

**For Fun and Prophet**
@BestOfBotWorlds
⚙  **Follow**

So Steve the 🐬 hid from Leonardo the 🎭 in the darkest corner of Cupertino

telling us what was said. But a bot that writes its own lines can focus on the negative talking points of the target, to craft an apt metaphor that is also a humiliatingly accurate put-down:

**MetaphorIsMyBusiness**
@MetaphorMagnet          ☼  👤+ Follow

But Richard the 🐾 humiliated Frank the 🐻 by calling the sociopathic and ruthless 🐻 the Keyser Söze of wielding political power

The dramatic irony of comparing Frank Underwood to Keyser Söze of *The Usual Suspects* (spoiler alert: Söze was also portrayed by Kevin Spacey— *or was he?*) is not beyond the reach of the metaphor generation process, as the NOC list allows just this kind of metaknowledge to be used when forming similarity judgments.[29] However, the choice of comparison above must rank as another happy accident in the mold of Nixon's snake. The bot, via its characterization of Nixon, does not intend to break the fourth wall, but that is the result nonetheless. When so many of a bot's choices are informed by knowledge, it becomes hard to tell when it is knowingly winking at its audience, though the larger point here is that any storyteller who pursues a knowledge-based approach to character formation is freed from a dependence on baked-in gag lines for its speech-acts. And just as one speech-act often begets another in human interaction, we might expect the butt of one put-down to be the originator of the next. The plot dictates that Underwood now hates Nixon for his temerity, but his insult is internalized:

**MetaphorIsMyBusiness**
@MetaphorMagnet          ☼  👤+ Follow

So Frank the vindictive 🐻 hated Richard the 🐾 for being jowly, deceitful and secretive

So the bot reaches into its Negative Talking Points for Nixon to pull out "secretive," "deceptive," and ... "jowly"? This may not seem like the most rational response, but in the bot's defense, an emotion as extreme as hatred is rarely rational. We humans also reach for the first pejoratives to mind when we lash out at others, and our bots—in their simplicity—have a tendency to mirror our least flattering features. But let's skip ahead to the end of this tale, passing over Underwood and Nixon's temporary rapprochement and subsequent falling out (again). The final action in the story has Underwood *cheat* Nixon, and this is rendered as a financial swindle:

**MetaphorIsMyBusiness**
@MetaphorMagnet                                                              ☼  ﹢ Follow

The closing bookend is perhaps more interesting, if only because it resonates so well—in what is not so much a happy ending as another happy accident—with our understanding of Underwood's character in his Netflix drama *House of Cards*:

**MetaphorIsMyBusiness**
@MetaphorMagnet                                                              ☼  ﹢ Follow

Bots may be clockwork contrivances, but they contrive *for us*, to create a series of happy accidents for our amusement and occasional incomprehension. We wind them up and set them loose so they might turn words and ideas into playthings and thereby wend their way into our imaginations.

### Toy Story Ad Finitum

Children love to play with dolls, and so their paraphernalia (sold separately) have become the crack cocaine of the toy industry, especially when the merchandise is shifted as part of a tie-in deal with a hit movie. Once a child becomes the proud owner of a Han or a Rey figure, Chewbacca and Leia and Luke and Darth become clear objects of desire, too, as do the scale-model sand speeders, TIE-fighters, X-wings, Millennium Falcons, and anything else that can be molded in plastic. But if children show a laser-like focus on the latest tie-in products in the run-up to Christmas, the story is very different *after* Christmas, once the packaging is cleared away and the kids settle down to some serious playtime.

There are no genre boundaries or franchise restrictions in the toy box, and children show an ecumenical zeal in the ways they play with toys and accessories from multiple franchises, even when those elements have vastly mismatched scales. A Barbie doll or a Disney princess can stand in for Princess Leia in a pinch, and a soccer ball makes a decent Death Star. George Lucas's hippy-dippy notions of "the Force" feel right at home in Hogwarts, so Obi-Wan Kenobi and Hermione Granger can make a great tag team against Darth Vader and Lord Voldemort (who makes an ideal Sith lord). Lego men and GI Joes can exist side by side, with a little Swiftian fantasy providing the necessary glue. Wittgenstein suggested that philosophers can learn a lot by watching children play: "I can well understand why children love sand," he said, but he could just as well have been talking about how kids play with any kind of toy with rich affordances to explore.

A child's imagination is rarely contained by anything so prosaic as the line between reality and fiction. Kids had Spider-Man square off against Superman in epic toy battles long before Marvel and DC got their acts together with a comic book crossover in 1975, and when DC pitted Superman against Muhammad Ali in 1978, it was long after kids had first put the pair on the same imaginary fight card.[30] Kids have fertile imaginations when it comes to inventing bizarre mashups and face-offs that cross conventional boundaries of time, genre, medium, and historicity. Hollywood has thus sought to foster a childlike imagination when appealing to kids with blended offerings such as 1943's *Frankenstein Meets the Wolfman*, yet as memorably satirized in Robert Altman's film *The Player*, many films aimed at adults are similarly motivated by cross-genre blends. So who can blame writers for wanting to make sport of their own gimmicks, as in this exchange in *Jurassic Park* that wittily exposes the cut-up at the movie's heart:[31]

**John Hammond:** All major theme parks have delays. When they opened Disneyland in 1956, nothing worked!

**Ian Malcolm:** Yeah, but, John, if the Pirates of the Caribbean breaks down, the pirates don't eat the tourists.

*Jurassic Park* is as childlike a blend (in the best sense of "childlike") as *Ali versus Superman or King Kong* versus *Godzilla* or *Abbott and Costello Meet Frankenstein* or *The Towering Inferno* (a film adapted from two novels, *The Tower* and *The Glass Inferno*) or any other mashup of narratives that you care to mention, from big-budget blockbusters to obscure fan-fiction blogs. This enthusiasm for coloring outside the lines has also given us TV's *Community, The League of Extraordinary Gentlemen, The Cabin in the Woods, Iron Sky, Penny Dreadful,* and the BBC's *Dickensian,* a show that throws all of Dickens into a blender so that Bob Cratchit can be arrested for the murder of Jacob Marley by Inspector Bucket of *Bleak House.* As we have seen in this chapter, our bots can play this game too, and play it well, for our amusement if not theirs. So the big lesson we draw here concerns the knowledge representations we give our Twitterbots. Real children must make do with imagination when toys are in short supply, but the more diverse the toy box that we can gift to our digital children, the more imagination they can show when playing genre-bending games for themselves.

## Trace Elements

Squeezing a whole story into a single tweet can be harder than squeezing a ship into a bottle. Yet we shouldn't overly concern ourselves with size limits, especially as far as Tracery and CBDQ are concerned, because the latter will not tweet outputs that exceed Twitter's size limits. You will find a pair of Tracery grammars for generating one-tweet stories in a directory of our TraceElements repository named *What-If Generator.* Each translates the causal structures of Scéalextric into simple grammar rules that generate the next state of a story (the right-hand side of the rule) from its current state (the left-hand side). The following is a tweet in which a two-act story fits within Twitter's original character limit:

What if Jaime Lannister was commanded by Professor James Moriarty but our "soldier" then disagreed with this "general"?

And here's a grammar output that requires the new 280-character limit:

What if Orson Welles translated for Tom Hanks and our "interpreter" was then trusted by this "listener," but our "intimate" then manipulated this "confidante," and our "cheater" then profited from this "sucker"?

In either case, notice how each action of the story uses metaphors rather than pronouns to refer to the participants of a previous action. Our context-free grammars have no memory of what has gone before, so these stories have no persistent memories of their protagonists, their names, or their genders. Yet the grammar rules that generate subsequent actions from current actions can use knowledge of the current action (the verb, not its participants) to generate referring metaphors for the participants of the next. This version of the grammar is named *What-if grammar backward.txt* because the referring metaphors always refer back to the semantics of the previous action. Another variant, called *What-if grammar forward.txt*, generates referring metaphors that are specific to the next action only. Try both in CBDQ to see which generates the most coherent narratives for you.

With a little help from CBDQ, we can also use Tracery to generate stories that extend over an arbitrary number of threaded tweets. The key is CBDQ's support for a response grammar, which allows a Tracery-based bot to respond to mentions from other Twitter users. If the core Tracery grammar generates the first act of the story *and* mentions itself in that first tweet, then the bot's response grammar can respond to this first tweet—in effect, respond to itself—with a follow-up action in a new tweet. If this follow-up tweet also mentions the bot's own handle, the response grammar will again be allowed to respond to itself with subsequent actions in subsequent tweets. This call-and-response structure marshals the two grammar components of CBDQ to allow a bot to generate a long-form story by talking to itself. You will find grammars for each side of the conversation in a directory named *Story Generator* in our TraceElements repository. You may notice that these grammars give names (such as Flotsam and Jetsam, or Donald and Hillary) to the characters in each story, and consistently use the same names for *A* and *B* across tweets in the same narrative. Different narratives may use different character names, so how does the grammar remember which names to use in different tweets?

We use another trick to build a long-term view into a grammar that lacks even a short-term memory. Rather than use the nonterminals of the grammar to represent simple story-states that correspond to plot actions, we create composite states that bind a plot action to the final action of a story. Thus, the grammar uses states such as *fall_in_love_with/are_betrayed_by* (which can be read as: *A* falls in love with *B*, but is eventually betrayed by *B*), and uses its rules to interlink states that end with the same final action. Because each story state "knows" how its story will end, it can use this knowledge to assign coherent character names across tweets. Thus, for example, stories that end with *are_betrayed_by* always use the names Mia and Woody. Since the grammar generates stories that terminate with more than two hundred unique actions, it uses a corresponding number of name pairs to name its characters in the same number of story families.

Incidentally, this strategy resolves another issue with grammar-generated stories, which have a tendency to pursue meandering and looping routes through their possibility spaces. These complex states ensure that stories approach their conclusions with a sense of momentum, and they also allow the grammar to know when to end a tale. A story that reaches a state $N/N$, such as *are_betrayed_by/are_betrayed_by*, will have naturally reached its predestined conclusion and have nowhere else to go.

# 9 Meet the Bot Makers

## Welcome to Botopolis

The Victoria and Albert Museum is tucked away in London's South Kensington district, nestled between the glittery lights of Harrods and Knightsbridge on one side and the solemn edifice of the Royal Albert Hall on the other. The area between the museum and the Royal Albert Hall makes up the Albertopolis, an area brought to life in the mid-nineteenth century by a huge wave of investment from Queen Victoria and Prince Albert, following a hugely successful public exhibition (called, imaginatively, the Great Exhibition) that was held in nearby Hyde Park. The profits from the Great Exhibition were set aside for huge investment into this area of London that is now bustling with landmarks, such as the Science Museum, the Natural History Museum, the Royal Albert Hall and Victoria and Albert Museum, the Royal Colleges of Art and Music, Imperial College, and more. All the museums in Albertopolis are free to enter, and the Royal Albert Hall runs countless free or inexpensive events, while Imperial College runs an annual science festival to showcase their research for the public. This is a part of London where art and technology collide with the public in the nicest of ways, so where better to meet and talk about Twitterbots? In April 2016 the Bot Summit, an annual day of talks and thinking about bots, came to Europe for the first time, hosted by the Victoria and Albert Museum.[1] The event attracted bot makers from all over the United Kingdom and many from overseas, thanks to a fundraising effort by the bot community to bring speakers from abroad. Any who couldn't make it in person were able to tune in to a live stream, which you can still find and watch online. Its organizer, bot builder Darius Kazemi, was there in person to coordinate the day, at the center of a melting pot for culture and science.

As we've seen throughout this book, Twitterbot creation is a curious mix of art and science—a "poetical science," to use the words of Countess Ada

Lovelace—and its various collectives have a lot in common with both artistic movements and scientific communities. In this chapter, we look at the people who make bots, how their communities have grown, and how the bot makers and their bots have influenced one another. As we'll see, there really is no single stereotype that encapsulates who a bot maker is or who a bot maker should be. Instead, it's a wide-open community where everyone is welcome and everyone can bring their own vision of what a bot can be and mold it into something new. Everyone is free to bring their own voices to the software they make and to the bots they create, and send something out into the world to tweet in their place.

Finding the Bot Summit was itself an adventure. Past several grand halls in the museum, up a marble staircase, and through several halls packed with display cases filled with inherited, gifted, or otherwise acquired works of art and historical artifacts, a small antechamber slowly fills up with unassuming figures. The only hint that there might be botters present is the occasional computer flipped open on a lap or on a nearby table. The lineup is a mix of young and old, smart and casual, quiet and boisterous. At the end of a small corridor leading out from the antechamber is a doorway that opens into a darker room packed with neat rows of chairs, where a beaming Darius Kazemi is busying trying to get another beamer to work.

The talks at the summit cover a huge range of topics, including celebrations of tools and techniques that make bot making more accessible, frank discussions of the politics of Twitterbots and their direct technological predecessors, and lively glimpses behind the curtain of some of the weirdest and most magical bot projects. This mix of topics and ideas, which do more than focus on the engineering how-tos of Twitterbot building, is a fitting structure for the Bot Summit, reflecting how diverse and thoughtful the world of botting can be. The live stream allows bot builders from across the world to take part in the event, commenting and chatting with one another and joining in the discussion and reflection around each talk.

In this chapter, we meet some of the people behind the Twitterbots and look at the community they have created. Like the Bot Summit itself, the bot-making community is a world of people of diverse ages, skill sets, and backgrounds who are interested in thinking about the broader implications of their work and their art and about the impact it can have on the world. We'll see how these concerns have shaped the kinds of Twitterbots that people build and how they have given rise to exciting new tools, websites, and organizations that intersect with and reach far beyond the world of Twitter.

## Allies and Alliances

The Twitterbot community is a vast and intricate web of people that stretches around the world, touching all manner of disciplines, backgrounds, and interests. Some people work happily in a single area, sharing their work with just a few friends, while others spread themselves across multiple technical and creative boundaries, to connect with large groups of people. Sometimes these people coalesce into communities, like the loose collective of bot makers who go by the hashtag *#botALLY* on Twitter (the pronunciation of this term is something of a mystery, with some preferring it to rhyme it with *totally* and others declaring themselves "bot allies" instead). Though many more work outside these groups, they all form part of the joyous mix of ideas and projects that make up the Twitterbot community. It is the *#botALLY* community that is responsible for organizing the Bot Summit each year, although it is less a fixed group with members and more a label that one attaches to oneself. The hashtag covers a mix of people asking for help as well as offering it. It unites those showing off their bots and those promoting the work of others, and it links those looking for creative collaborators to those offering resources for everyone else. Because the community is so widely distributed and because social media by definition can make us feel both connected and more isolated, many bot makers do not consider themselves officially part of the *#botALLY* community. But most have been touched or influenced in some way by its members through the resources they make, the philosophy they promote, or the bots they build.

The interconnectedness and breadth of the bot-making community is its most defining feature, a point that is often raised by bot makers when they are asked about what ties them together. "The thing I appreciate most about the community is that there's room for everything and everyone" says Ashur Cabrera, a Parisian software developer and bot maker, "You'll find broad support and encouragement among bot makers for you to bring your brilliant, weird, funny, tiny, sprawling, somber, silly, activist, emoji-only idea to life." Bilgé Kimyonok, another bot maker from Paris, says this diversity carries through to their bot creations too: "I think anybody can find a Twitterbot that suits them, or at least interesting enough to be read and followed." This diversity stems in large part from the efforts of community members to be positive, diligent, and active in promoting diversity and making the bot space welcoming to everyone. As a bot maker known enigmatically as "the Doctor" puts it, "The Twitterbot community made a lot of mistakes and learned from them; that's why we have a collection of best practices that we follow."

Another reason for this huge diversity is the intriguing mix of the creative and the technical skills inherent in the building of any Twitterbot that expresses a creative idea in program code. Many bot makers straddle this gap between art and technology with ease, with one of the best-known examples of a bot builder with a foot in each domain being Allison Parrish, the mind behind the @everyword bot. Allison is the digital creative writer-in-residence at Fordham University, a job description that already hints at a love of cross-disciplinary work. She has been a software architect, a researcher, a chief technology officer, and a poet, and bot making allows her to combine all of these efforts into one singular activity.

"I honestly believe that people are hungry for poetry," she explains, "hungry for language arranged in unusual ways, language that challenges your ability to read it in conventional ways." Poetry is a defining theme for Allison: while @everyword is her best-known bot, and perhaps one of the best-known bots ever created, she has many other bots that play with the poetry of language and algorithms. One particularly beautiful bot that we have encountered already in this book is @the_ephemerides, a Twitterbot that juxtaposes computer-generated poetry with photographs of alien worlds plucked from space probes drifting through the solar system. The results are haunting, combining the alien confusion of machine-generated poetry with the quite literally otherworldly atmosphere of outer-space photography. The tweets find themselves nestled in your time line amid GIFs of small animals eating things they shouldn't and Promoted Tweets about self-help guides.

"Twitterbots are very 'usable' in the Donald Norman sense of the word," Allison explains. Norman is an academic, a psychologist, and a designer who wrote *The Design of Everyday Things*, which advocated for the kind of design in which usability is a joyful, beautiful process that makes a product feel natural and makes the user feel good.[2] She adds, "[Twitterbots] post small chunks of text that can facilitate deep engagement but don't demand it, which is a perfect format for poetry." A more recent example of this is Parrish's @a_travel_bot, which posts excerpts from imagined travel guides to fictional places. Each tweet has a heading in the style of a travel guide chapter, with some additional text to offer information of the corresponding kind for this imaginary place. "HERITAGE AND STEAM RAILWAYS," one tweet begins. "It runs for seven miles through scenic hill country. It is the longest heritage railway in the country." The bot veers between plausible texts like this one, a mashup of very real locations with places that are fantastical and surreal, all mashed together into 140 characters or fewer. "UNDERSTAND," another tweet explains. "It is not possible for you to get lost. During the summer, the parties and mosquitoes enjoy late hours."

Parrish is very fond of @a_travel_bot, a key reason being that it departs from the emotional tone more commonly taken by other linguistic bots. "Lately I've been interested in trying to expand the range of emotions and experiences that my procedural writing evokes," she says, because "there's a sort of gentle absurdity and meditativeness to the text that the bot generates, which is a good feeling to get from procedural writing." Her @a_travel_ bot also demonstrates a canny ability that Parrish shares with many other bot makers—an ability to seek out clever and elegant corpora full of data and to think of the perfect procedural system to build on top of them. The structured language and common patter of travel guides resonate with us, allowing the slightly weird combinations of places and ideas to feel a little less alien and a little more relatable.

While a great deal of the bot-building community's data is taken from public repositories of information such as WikiHow, Wikipedia, and even Twitter itself, the community has also come together to build its own resources for bot makers to reuse. For instance, Darius Kazemi has put together Corpora, a collection of useful and interesting data sets that are preprocessed to be clean and ready for use in bots, or indeed in any other kind of generative endeavor.[3] It is an inventory that we might imagine a character in a Borges story compiling and includes such things as a list of jobs, a list of the names of Fortune 500 companies, a list of UK political parties, a complete list all of the Greek titans, a list of Christian saints, and an index of Antarctic birds grouped by family. The eclectic mix of data is in part a result of the project's origins and the way it has been gradually pieced together by the community. Each data set is the work of someone who wants to make bot making a little bit easier for someone else.

One person with years of experience in assisting the bot-making community is Erin McKean, one of the founders of Reverb, the company that runs the Wordnik website. Wordnik is a special kind of online dictionary that prides itself on having one of the largest word lists in existence.[4] It voraciously gobbles up examples of words in use, using its own army of bots to scour the web for language being used in new ways so it can index these uses in its dictionary. This open attitude to language is what gives the Wordnik dictionary its unique flavor and expressiveness. Wordnik is always waiting to welcome new language into its lists in its rawest forms. But it is more than just a dictionary: it is a rich store of images, synonyms, use cases, concepts, and other miscellaneous items. It is also a good friend to the Twitterbot community. Wordnik has an open API, which many bot makers use, so Erin has frequent interactions with the #botALLY community. Wordnik has even sponsored parts of the Twitterbot community, including Bot Summits like the one that took place in London in 2016. "They're creating gifts to the world," Erin says of bot makers.

Erin is a programmer and a lexicographer, so playing with language in both its artificial programming and natural spoken forms is something that has been on her mind for many years. She recalls making bots as early as 2008, using a service called Yahoo! Pipes that allowed developers to plug different data sources together to make simple web apps. These days, her favorite creation is a personal one: @adoptnik, a Twitterbot that tells the world every time someone adopts a word on Wordnik. Make a small donation to keep the site running, and Wordnik lets you take in a lonely word and become its adopted patron, displaying your name on that word's online page.

### First, Do No Harm

When Erin talks about Twitterbots, she conveys a feeling of immense positivity that is shared by many people in and around the bot community, a feeling that these odd creations are not just interesting pieces of software but positive forces in people's everyday lives. She argues that "[Twitterbots] take you outside your news-outrage-and-sandwiches timeline to give you a minute to look at the world in a new way. ... I really appreciate that the people in the community stop and think about the effects their bots have on others—just because you can do something, doesn't mean that you should." Erin is referring to the particular focus that many members of the community put on ethical bot making. Many bot designers have weighed in on this topic, including Darius Kazemi, who formalized a set of guidelines for ethical design that we discussed back in chapter 2. For some bot builders, these guidelines can seem overly cautious, but for Darius, this is the point. Staying a safe distance from the boundaries is important when there may be serious consequences for crossing the line *even once*.

His guidelines are obvious ways to avoid annoying people and not getting banned from Twitter, but of course there are many other ethical complications to do with Twitter that are not covered by the company's spam policy. Indeed, Twitter has earned a reputation for being less than stellar at curtailing the behavior of the platform's worst offenders, both human and digital. So bot builders like Darius try to take things a step further in avoiding giving offense or ever causing hurt with a bot by creating resources that help builders develop bots that can be that much more intelligent and self-aware (or self-limiting) about what they do on Twitter. One example is his *wordfilter*, a list of words that are likely to cause offense or have their roots in slurs and insults.[5] Had Microsoft used a similar list, or indeed this specific list, @*TayAndYou* might have recognized the offensiveness of nasty words in the bot-baiting tweets of malicious users and avoided reusing them itself. Many Twitterbots use data from public sources or other tweets written by human beings, and by putting that text out into the world using their own voices, they are, like @*TayAndYou*, potentially repeating odious things. As Darius put it in an interview, "I don't want my bot to say anything I personally wouldn't say to a stranger."[6] His wordfilter currently contains sixty-six words personally added and checked by Darius. This doesn't include obvious swear words that lack any personal basis, such as the scatological words, as he puts it. But again, Darius is more interested in staying a safe distance from danger, and his list has been intentionally designed to be overzealous in filtering out dubious words and removing any lexical items that begin with potentially nasty prefixes. "New slang pops up all the time using compound words, and I can't possibly keep up with it," he explains, so "I'm willing to lose a few words like 'homogeneous' and 'Pakistan' [from bot tweets] in order to avoid false negatives."

A false negative means that a reader may be hurt or offended by something that has slipped through the net, whereas the false positives that arise from the zealous removal of words that *might* potentially sound bad (but probably don't in any given context) make it that much harder for our bots to temporarily go over to the dark side. In the worst case, you simply run your generator again and make something new; that's the real beauty of writing a piece of software that can make other things. There's always something new to be made, and sacrificing a few extra outputs to be a good digital citizen is a laudable trade-off.

At times, this approach can require a great deal of nuance and insight. One of Darius's best-known bots, @*twoheadlines*, originally had problems when it mixed headlines about people of different gender identities. An early version of the system might have tweeted a mashup such as "Bruce Willis Looks Stunning in Her Red Carpet Dress." But a tweet like this posed a problem for Darius. On one hand, there's no reason at all why someone named Bruce Willis might not wear a dress and choose to use the pronoun "her." On the other hand, a common interpretation might be that the tweet implies that a man wearing a dress is innately funny, or that it is innately pejorative and innately humorous to show a man as having a woman's qualities. Again, Darius adopts a cautious attitude and has given his bot the sensitivity to stay away from issues such as gender mismatch, even if it inadvertently reinforces a gender-binary mind-set in the tweets of *@twoheadlines*.

Even with these precautions, bot ethics and etiquette are evolving concepts that are continuously explored, reformed, extended, and tested by practitioners, and much of this work and the discussions centered around them take place in and around the *#botALLY* community. To see what happens when creators fail to tackle these issues properly, we need look no further than Microsoft's *@TayandYou*, which we discussed earlier in this book. While Tay was undoubtedly a sophisticated project for Microsoft, it was also one with glaring errors in its design. Its attempt to learn from its human followers led it to rapidly learn from the worst kind of stimuli, to the point where it was soon parroting the racist and sexist language of bigots, endorsing inappropriate photographs, and generally promoting whatever dubious ideas it was told to promote. Microsoft was forced to shut the bot down within a day of it going live, with Tay explaining that it needed some sleep.

The problems inherent in Tay had nothing to do with any lack of resources or programming talent at Microsoft. Rather, what Microsoft's developers seem to have lacked is an understanding of the environment that bots exist within and of whom they interact with, of what they should be doing and of how they might be playfully subverted by others. Without properly understanding the social and ethical factors that have been the subject of much reflection by the bot-building community, it is all too easy to stumble badly when releasing a bot out into the wide, wild world of Twitter. Any bot that relies for its subject content on an unknowable body of external texts is vulnerable to malicious subversion, but when the bot also serves as the digital flagship for a global brand, the personal embarrassment that sabotage can bring can quickly turn to global humiliation. So it is no surprise that when corporations seek to promote their brands with bots or other social media interventions, they put a tremendous amount of thought into the kinds of language they will tolerate as input to, and potential output from, their online playthings.

When Coca-Cola launched its *GIF the Feeling* promotion, which invited consumers to attach a slogan of their own making to a Coke-themed animated GIF (showing antics that would not be out of place in a Coke TV spot), the people involved spent a great deal of time and energy imagining the kinds of language they did *not* want associated with the product.[7] Writing in the *Atlantic* magazine in January 2016, Ian Bogost described his attempts to reverse-engineer the profanity filter used by Coke's GIF app to prohibit the generation of brand-tarnishing memes with the potential to go viral.[8] Bogost ran a whole dictionary against the app, to see which words would cause it to balk, and was surprised by the results. Not only were all the usual suspects on the no-fly list, as well as a slew of words that might be used to craft an anticapitalist message (such as "capitalism"), so were a great many other words, from "couscous" to "igloo" to "taco." We can only imagine that the company was worried, with reasonable cause, that some users might use the connotative power of these words to create ethnic or racial slurs. Only a year earlier, Coca-Cola had launched a Twitterbot that would repurpose the text of a tweet bearing the hashtag #*MakeItHappy* as a work of ASCII art in the shape of, for example, a cutesy balloon animal. The company was to learn the hard way that ugly online texts do not require any obvious profanity. To prove the point, the website Gawker had no trouble getting the bot to make ASCII balloon animals from the slogan of the white supremacy movement and retweet it with the official company handle, @*CocaCola*. To hammer the point home, Gawker's Adam Pash went on to tweet the contents of Hitler's *Mein Kampf* at the bot, attaching Coke's innocuous hashtag to the extracts.[9] The Twitterbot's simplistic profanity filter saw no reason at all to prohibit the text, which was promptly repurposed as a giant happy face and retweeted. So Coke can be forgiven for its zeal in tagging innocent-sounding words as potential troublemakers, though it really is a mug's game to attempt to predict how others might use words (and their spellings) in creative ways. If in any doubt about this, just ask a spammer.

The subject of ethics for bots, much like that of ethics for humans, is a constantly moving target that elicits different opinions from different people. From Cokebot's gaffes to Tay's indiscretions, making software that heads out into the world to speak and perform is much more than a wholly technological challenge. As you work on your own bot creations, you'll discover both familiar problems that many developers have tackled before and a host of unique issues that only you will be able to solve. The first rule is not a moral injunction but an obvious product of common sense: if your bot is a text mill that takes its grist from the texts of others, be choosy about where those texts are coming from. Your bot may take its ingredients from a respected provider of news content, or a well-policed source such as Wikipedia, or from a corpus of nineteenth-century poetry, or from the outputs of a government-funded agency such as NASA, but you are asking for trouble if it draws its content from anonymous strangers on the Internet. The world of bot making is still an inchoate one, and there is much ground yet to be covered. The personal nature of these ethical dilemmas means that in some cases. you might feel the need to be more cautious, and at other times you may feel like pushing the received limits. Nonetheless, whatever you do, a little common sense can go a very long way when building a respectful bot.

Earlier in this book, we discussed Leonard Richardson's cautionary appeal to bot builders: "Always punch down." A strong theme throughout the Twitterbot community is thoughtfulness about one's work and about the potential impacts on the wider Twitter community. This is as good and powerful a reason as any to think carefully about what direction you punch in. But if that's not enough for you, Richardson offers another compelling reason: the web is a mess of legal situations and complications, and should your Twitterbot ever get you into trouble regarding slander, copyright, terms of service, or worse, you may need all the public support and sympathy you can get.

## Prêt À Tweeter

The community spirit that drives the Twitterbot community goes beyond bot makers to include organizations like Wordnik, which primarily works on technology and ideas adjacent to the Twitterbot sphere. This spirit is also what drives the bot community's desire to give back to others, in ways that go beyond simply making and releasing another bot. As we saw in chapter 3, Cheap Bots Done Quick is an excellent example of the way the bot-making community sets out to be open and welcoming, beyond simply sending a fresh little Twitterbot out into the world every other day.[10] Run by George Buckenham and powered by the Tracery grammar system designed by Kate Compton, CBDQ aims to make the creation of simple Twitterbots something that anyone can do easily. This desire to make bot making accessible to as many people as possible is one that runs through the entire community. Allison Parrish believes this desire is fundamental to the strength of the bot-making world: "Bot makers run the gamut from being computer science wizards to being completely uninterested in programming as a practice," she says, "and that broad range of interests and attitudes deeply enriches the kind of work that comes from the community."

The ideas behind Tracery began in 2005 when Kate Compton built a story generator as a project for an Interactive Narrative class. It wasn't until late 2013, when studying for her PhD, that Kate began fleshing out the ideas into a tool that anyone could use to build simple generators. Kate

Kazemi notes, for example, that observing repetitive behavior on the web, such as bad jokes, to which he attributes the origins of *@twoheadlines*, is the genesis of many of his projects. This kind of repetitive behavior also forms a major part of online activism and fuels discussions about social justice, all of which prompted Stanton to create *@101atron*, a bot that helps people automatically dispense informative links to people who ask well-meaning but somewhat trite questions, such as, "What is feminism?" Because these questions are often repeatedly asked of people who are already under the pressures of prejudice and oppression, the *@101atron* bot helps to relieve some of that pressure by cataloging links and serving them up to the people who ask the questions.

Stanton and Kazemi extended the ideas behind *@101atron* by working with activist DeRay Mckesson, in a collaboration that resulted in the creation of a bot named *@StayWokeBot*. The new bot has a broad list of topics that it can talk about and an improved mechanism for remembering who has said what to whom. It was also friendlier and sent a little machine-generated poem to every person who followed it. Yet the bot was more than a mere technical development, since Feel Train also improved the way that people can update the bot so that it can be easily maintained by someone who is lacking in programming skills. Many of the Twitterbots we have discussed in this book are single-use systems that are made by a single person to be put out into the world and then left to their own devices. *@StayWokeBot* is an example of a sophisticated bot that has a bigger role to play in the world than promoting the ideas of a single person.

The Twitterbot community is much more than a collection of grassroots movements and loose alliances of artists and engineers. While there may be few opportunities to make money from bot making right now, public and media interest in AI is reaching a point where Twitterbots become an obvious focus of interest for technology companies and their researchers. In this chapter, we have already seen one such example of this interest, in our discussion of Microsoft's abortive attempt to explore the world of bot-making with its well-meaning but easily misled Tay. For just as bots can amplify and multiply the ingenuity and creativity of a single person, these companies and organizations hope that a bot can similarly amplify and multiply the reach of their slogans and ad campaigns.

These commercial bots also fit snugly into the general categories we identified in chapter 2, even if these categories offer the kind of classification that we might be more inclined to associate with creative and artistic software. Bots such Bank of America's *@BofA_Help* search Twitter for specific words and phrases so they might interact with potential customers, in much the same way that a bot like *@StealthMountain* looks for users who have unwittingly misspelled "sneak peek." While the motives of *@BofA_Help* are not as mischievous as *@StealthMountain* and its ilk, it makes exactly the same kind of mistakes as those other bots. For instance, the account has responded to people discussing protests outside banks, foisted itself into conversations between other Twitterbots, and happily responds to insults with an adorable manifestation of its ignorance. When one user demanded, "Why don't you stop being pure evil?" the bot cheerily replied, "We'd be happy to review your account with you to discuss any concerns."

Other promotional and consumer-facing bots wait for others to interact with them, so as to provide users with something in return, in a mode of interaction that is similar to that of image-processing bots like *@LowPolyBot* and *@BadPNG*, or indeed of bots like *@StayWokeBot*. In 2014, the sportswear maker Puma ran a promotion around its Forever Faster brand, in which users were encouraged to tweet their names along with a hashtag asking to be included in an automatically generated message from a celebrity athlete. But without thorough filtering or due consideration as to how the campaign might be subverted, mischievous users were able to rapidly find ways to undermine the system, leading to heartfelt messages—apparently from Puma's celebrity spokespersons—thanking "cocaine" (and other, less printable benefactors) for their support and help in achieving their goals. Oops.

While bots like Puma's digital image maker offer new ways for the company to communicate with others, bots such as *@BofA_Help* might seem to represent a colder and more troubling mode of company interaction, as there seems to be an express interest in supplanting human labor with a purely digital workforce. The designers of corporate bots such as these set out to use the scripts and templates that human customer service workers already follow by translating them into program code that can do away with the need for a human in the loop. Unlike the artistic bots that magnify a person's labor, such bots diminish the role of humans. As a society, we are rightly concerned that autonomous AIs might be coming to take our jobs, but rarely does our anger and suspicion actually fall at the feet of the parties behind this drive. For now, corporate bots seem charmingly inept at their jobs and remind us why we value having a real person in that loop after all.

Automating social media interaction via bots may be appealing for companies, but it also belies a certain lack of understanding about how social media actually work and of how people tend to treat machines differently from people when they interact with them. Companies clearly feel the wonderful people who make up the bot community, from independent creators working feverishly on passion projects to large collectives and companies all pulling together on big ideas. It's a web of people connected to each other and to each other's creations. If you're reading this book and experimenting with Twitterbots for the first time, the wider bot community might seem like a somewhat daunting mix of experienced old hands, close friends, big corporations, and a sea of hundreds of millions of Twitter accounts to get lost in. But almost all bots start in the same way: each is ushered into the world gingerly by the people who created it, shared with people they know, to slowly create things that other people are drawn to, delighted by, and eager to share with others.

Hashtags like *#botALLY* are there for all of you who want to reach out to a larger collective, but every bot creates a new community around it, whether it's a burgeoning mass of tens of thousands of people, or just your four best friends who enjoy playing with something that you have created. One bot maker named Glen nicely describes Twitterbots as "art, once removed." As he puts it, "The artist comes up with the concept, and then steps back. ... It's art, but created by creating an environment, like growing a plant." Every bot is a product of the people around it: those who inspired it, those who built it, those who followed it, and those who were inspired by it. Each one starts with someone like you, deciding to make something that makes something.

## Trace Elements

The inspiration for a new bot can come from many sources. Some are inspired by a bot builder's drive toward social activism, others by the social conscience of a partner, a patron, or a business client. Some are inspired by the news or a conceit from a movie or a game, and others are driven by the creative opportunism that arises whenever a builder stumbles upon a useful new web service or a marvelous trove of data. We conclude this chapter by considering one such data resource: the structured online database *dbpedia.org*

Unlike Wikipedia, the source from which it derives its content, dbpedia is a knowledge base not of free-text articles but of semantic triples. Stored in a series of large text files that can be downloaded from the site, its triples are easily extracted on a one-per-line basis. One text file contains categorization triples that link specific instances of movies, games, and people to informative semantic categories. Another contains taxonomic relationships that link these categories to one another via *broader_than* and *narrower_than* relations. These categories often have an interesting linguistic structure of their own. For instance, the film *Blade Runner* belongs to the categories *Films_about_altered_memories* and *Flying_cars_in_fiction*. A shallow parser that skims DBpedia's files can easily extract the facts that Blade Runner is a thematic mix of robots, genetic engineering, altered memories, and flying cars.

In a directory of the TraceElements repository named *DBpedia Riff Generator*, you will find a pair of Tracery grammars that exploit a large body of these easily mined dbpedia associations. Here is a representative output of the grammar named *DBpedia grammar.txt*:

After the film The Fly II, I dreamt of engineers who study genetic engineering and collect insects, @MovieDreamBot.

Notice how each tweet mentions the bot's own Twitter handle, prompting a corresponding response grammar (*DBpedia responses.txt*) to generate a reply:

Hey @BestOfBotWorlds, spin us a yarn about how some manufacturer hired this engineer.

The reply also mentions a Twitter handle, yet it is not the bot's own. Rather, it is the handle of another bot whose own response grammar is now prompted to generate a response. We saw in the previous chapter that a Tracery-based bot can generate a long-form story over a series of connected tweets by effectively conducting a conversation with itself. So what we see here is one bot throwing a conversation starter to another. Our dbpedia bot plucks an idea for a new story from the low-hanging fruits of a large public database, and passes this idea to another bot so it can be developed into an actual story. The resources that underpin a bot's behaviors are frequently data resources, and sometimes web services. But it is worth remembering that our bots can themselves be wonderful resources for the building of other bot behaviors.

## Alternative Facts

Fictional what-ifs come in all shapes and sizes. Many are ephemeral brain farts that are as disposable as they are whimsical. What if Neo had chosen the blue pill? What if plants could talk? What if animals *wanted* us to eat them? What if we needed stamps to send email? What if Steve Jobs was an alien who merely returned to his home planet? What if Donald Trump turned the Whitehouse into a casino? Other what-ifs, born of exactly the same slice'n' dice attitude to reality, turn out to be made of weightier stuff and provoke fascinating debates. What if animals brought a class action suit against the human race? What if the ancient Romans had invented the atom bomb? What if the USSR had won the cold war? What if the Axis powers had won World War II and divided up the United States among themselves? This last what-if is the provocative basis for Philip K. Dick's Hugo Award–winning novel, *The Man in the High Castle*.[1] Written in 1962 and set in the same year, the book explores an alternate reality in which the allies lost World War II after the Nazis dropped the first atomic bomb on Washington, DC. The political counterfactuals of the novel concern the machinations of Nazi Germany, which now controls the Eastern states of America, and of Japan, which controls the Pacific states, with the Rocky Mountain states serving as a neutral buffer.

But Dick's novel is, more than anything else, a philosophical inquiry into our tangled understanding of reality versus appearance, fact versus fiction, authenticity versus artifice, and fate versus chance. His fictional Japanese occupiers of the Pacific states exhibit a fetish-like desire for the collectible vestiges of the prewar United States, collecting anything from old bottle caps to weapons and furnishings. Skilled counterfeiters realize tremendous profits in this seller's market, and Dick's novel follows both a purveyor of high-price antiquities and a producer of high-quality forgeries who later branches out into the creation of original items of contemporary jewelry. Dick uses these complementary perspectives to explore the inherent value of an artifact. What does it mean to say that something is a "fake" or the product of artifice? Does an object that "has history in it," such as an object tied to a famous person or a pivotal event in history, have more intrinsic value than a perfectly functional copy of the same thing? Does the origin of a thing, or the intent with which it was made, wholly determine its usefulness to others? This is a novel in which fakes of all kinds abound, from people who are not who they seem to be or who they say they are, to competing histories that never completely persuade. The characters of the novel may hold differing views on "historicity," but each in his or her own way attempts to project external meaning onto objects and events to see their way to an "inner truth." In many ways, the themes of Dick's novel are as applicable to our appreciation of artifacts made by autonomous machines as they are to more conventional man-made artifacts, and perhaps all the more so because our machines are themselves a special kind of man-made object. We ask many of the same questions of their outputs and face many of the same doubts. In this final chapter, we shall find that many of the themes of Dick's novel chime with the opportunities and the concerns raised by our bots.

Consider the desire for *historicity*, that is, our need to find a connection between an artifact and some external context that can give it relevance and meaning; the search for history is ultimately a search for *story* and a desire to frame an object within a satisfying narrative. Dick's novel has a cynical manufacturer of forgeries make the case that historicity is a comforting story we tell ourselves and others. Holding up two identical-looking Zippo lighters, only one of which was owned by Franklin D. Roosevelt (who was assassinated prior to World War II in the time line of the novel) he argues, "One has historicity, a hell of a lot of it. As much as any object has ever had. And one has nothing. Can you feel it? ... You can't. You can't tell which is which. There's no 'mystical plasmic presence,' no 'aura' around it." An imitation Colt revolver fills the novel's antiquities dealer with shame after he has been duped into selling it, but the fake proves to be just as effective as the real thing at killing two black hats in a shoot-out redolent of America's Wild West. Yet we shudder at knockoffs and fakes even when they are made to exacting standards because their narrative is one of deception in which we are the dupes. Our bots do not subscribe to this narrative because they wear their artifice on the sleeves. We humans knowingly follow bots *because* they are bots, and not for any of the reasons that we buy knockoff products even when we know they are fakes. For we

of designations for human craftsmanship that he borrows from Eastern philosophy. The first is *wabi* (or *wabi-sabi*), a Japanese term that loosely translates as lean, spare, and graceful. An object has *wabi* if it lacks unnecessary frills and fulfills its function with a no-fuss, no-bullshit grace. An artifact with *wabi* will have earned its imperfections and will wear them well as a sign of its historicity. Due to its restriction on the length of tweets, Twitter seems an ideal place to look for the linguistic equivalent of *wabi*: a well-crafted news headline may have *wabi* in spades, as might a finely wrought joke that says no more than it absolutely must to achieve its humorous effect. The "novels in three lines" of modernist writer Félix Fénéon, whom we met in chapter 1, appear to have been constructed with *wabi* as their chief artistic motivation. *Wabi* abounds whenever Twitter is used masterfully as a medium, as in the *@novelsin3lines* account that was retrospectively created to showcase Fénéon's oeuvre in English, yet because the designation is a term of discernment, *wabi* is, sadly, far from the norm. The Japanese in Dick's novel are connoisseurs of *wabi* and eagerly seek it out in items of collectible Americana, but they cast a cold eye on anything that is new or lacking in history, or seemingly without a useful function that might offer a larger context in which to judge its *wabi*-ness. A forger of antiquities in the novel turns his hand to making artisan jewelry, and a dealer in antiquities takes some of the jewelry on contingency. The dealer, Robert Childan, presents one of the pieces, a decorative pin, to Paul, a valued Japanese customer, but he is not impressed with the offering. At first Paul is confused: the object, a shiny "blob" of polished metal, seems altogether risible, and his friends snigger in agreement. Though Paul is embarrassed for the dealer, he cannot stop himself from stealing glances at the pin, for reasons he does not yet understand. The object haunts his thoughts, and when Paul meets again with Childan, he offers these observations:

> It does not have *wabi*, Paul said, nor could it ever. But—He touched the pin with his nail. Robert, this object has *wu*.
>
> I believe you are right, Childan said, trying to recall what *wu* was; it was not a Japanese word—it was Chinese. Wisdom, he decided. Or comprehension. Anyhow, it was highly good.
>
> The hands of the artificer, Paul said, had *wu*, and allowed that *wu* to flow into this piece. Possibly he himself knows only that this piece satisfies. It is complete, Robert. By contemplating it, we gain more *wu* ourselves.

We do not need a larger frame of reference in which to appreciate *wu*: an object with *wu* is sufficient onto itself, exhibiting inner balance and harmony. Broadly speaking, whereas *wabi* is a quality found in man-made artifacts that satisfy their functional demands with unshowy elegance and grace, *wu* is a quality most often found in natural objects that have no designer or no predetermined function to serve. We might perceive *wu* in the symmetry of an image woven from a complex cellular automaton, though it is stretching the point somewhat to talk of *wabi* and *wu* as possibly inhering in linguistic tweets, especially in the tweets of an automated bot. Nonetheless, we gain a certain amount of leeway by having the term *bullshit* anchor the other end of our aesthetic spectrum, and while tweets hardly count as natural objects, it is no accident that we use the word *sublime* to describe both the ineffable wonders of nature and the wonders of a poetic turn of phrase. The most exquisitely wrought aphorism, for instance, combines a lightness of touch with the sense that one could not have said it better, as not a single word can be profitably changed. Such a phrase needs no historic frame of reference in which to be appreciated, save for the frames that unite us all: the frame of language and the frame of human existence. So the bon mots of Oscar Wilde and Dorothy Parker have inner harmony, complementarity, and balance in abundance, and to enjoy them is to feel that some *wu*-like quality has flowed straight from the writer into his or her words. Aphorisms such as these are self-contained and complete, and as Dick's character Paul suggests, by contemplating them, we gain more of their *wu*-like quality for ourselves.

Fortunately, *wabi* and *wu* are not all-or-nothing concepts; rather, they are a continuum along which our bots might gradually progress with time. Just as @*LostTesla*'s tweets are occasionally Zen-like, there is a certain *wu*-like self-sufficiency in the solipsistic metaphors of @*metaphorminute*, which showcases the exuberance of language without trying to mean anything at all, or in the way that @*NRA_tally*'s tweets counterbalance the actions of two different kinds of gun fanatic. To the extent that the tweets of any Twitterbot exhibit either *wabi*- or *wu*-like qualities, it is because the bot has been designed to embody those qualities, so that they might flow from builder to bot to tweet. The bot itself may thus exhibit *wabi* if it embodies a simple but appealing idea with leanness, spareness, and grace. As such, the possibility does exist for our bots to add to the collective *wabi* and *wu* of the Twittersphere with the products of their linguistic and visual invention.

## Consulting the Oracle

The most intriguing what-if in Dick's novel *The Man in the High Castle* is not his alternate history of an allied defeat in World War II: that conceit is as evergreen as the notion of using a time machine to assassinate Hitler or warn FDR about a Japanese attack on Pearl Harbor. No, the most counterintuitive counterfactual is Dick's suggestion that the people of his alternative time line, living under the totalitarian yoke of the Third Reich and Imperial Japan, would make the *I Ching*, the ancient Chinese system of oracular divination, an integral component of their everyday lives. Dick's characters consult the texts of the *I Ching* for insight into all of their moral questions, big or small, and it comes as naturally to them as tossing a coin, playing paper-rock-scissors, or reciting "eeny-meeny-miny-moe." But rather than give decision makers a binary random variable (whether *heads/tails*, *win/lose*, or *it/not-it*), the *I Ching* serves up a random signpost into a decision space of sixty-four possibilities, or *hexagrams*. Dick's characters generate the hexagrams of the *I Ching*, blocks of six lines apiece where each line is either solid or broken, giving $2^6 = 64$ possibilities, by throwing yarrow stalks or by tossing coins. They then look up an analysis for the hexagram that chance has given them in volumes of ancient commentaries, in a process called "consulting the Oracle." To understand how Dick's characters use the randomness of the *I Ching* to systematically weave a meaningful narrative around their actions is to understand how chance can be purposely harnessed by any decision-making agent, whether a human or a bot.

We all find ourselves blocked and stuck in a creative rut from time to time, and it is more easily said than done to look at our problem with fresh eyes. One way to force a new perspective upon ourselves—according to creativity mavens such as Edward de Bono– is to actively engage with a random but meaningful stimulus.[9] We might, for instance, open the dictionary at a random page, pick a word with our eyes closed, and then try to integrate one or more senses of this word into our thinking about our problem. Though utterly out of left field, these fresh elements may be just the stimuli we need to escape our rut. But we are not limited to the dictionary when we play this game, for we could just as easily use the Bible, the Quran, the Torah, the *Guinness Book of Records*, *Bartlett's Familiar Quotations*, Wikipedia, or the tweets of Donald Trump as our source of external stimulation. This simple method has a surprising provenance and is a whimsical update of an ancient practice called *bibliomancy* in which, for example, Christians looking for moral guidance might pick a seemingly random chapter and verse from the Bible in the hope that God, or providence, has guided the selection, just as a Muslim might do the same with the Quran. Of course, any randomly chosen text fragment is not an answer in itself, but if one believes that the selection has a divine mandate, then one will look all the harder to see its potential relevance. So when we strip away the veneer of mystical mumbo-jumbo from the *I Ching* and set aside the notion that it allows us to read "the tenor of the universe" at a given moment in time (as one of Dick's characters memorably puts it), what is left is an ancient version of de Bono's dictionary method of bibliomantic inspiration, albeit one that has attracted volumes of sage commentary from ancient philosophers. This is the real psychological value of the *I Ching*: it cleverly exploits randomness in a process of systematic self-examination. Its random stimuli may come from without, but the answers to our specific questions must still come from within.

Dick was turned on to the *I Ching* in 1961, a year before he wrote *The Man in the High Castle*, and by all accounts he took it rather seriously as a method of inspired decision making. Just as his characters frequently "consult the Oracle," and choose their actions to fit the hexagrams that they randomly generate, Dick generated the hexagrams for them at these junctures not by inventing what his plot required but by obeying the *I Ching* himself. That is, he would throw his own yarrow stalks to form his own hexagrams, which would then dictate significant aspects of the plot when they were integrated into the text as character actions. This does not seem so very different from how one might write a sword and sorcery novel by co-opting the dice-based mechanics of *Dungeons & Dragons* to choose among plot outcomes, and we can think of Dick as a high-brow *dungeon master*. [10] It is a testament to his discipline as a writer that his random D64 rolls sometimes took his story down avenues that Dick would have preferred not to pursue and cut off others that might have better reflected his desired shape for the story. Yet his use of randomness was not deterministic, as it is in simple D&D bots, because the *I Ching* is not deterministic: it uses randomness to engage subjective thought processes, not to determine the results of those processes. Nonetheless, by using randomness systematically, with a disciplined and almost algorithmic respect for the results of stochastic processes, Dick used an approach to plotting that is not so very different from the algorithmic storytelling of our what-if machines. A bot can likewise use randomness as a guide to decision making without being wholly determined in its actions by the results of random number generation. We have seen, for instance, how random outcomes might decide the high-level structure of a plot by determining the next triple of actions in a story arc, *and* how a bot may yet control how each of these actions is to be fleshed out using knowledge of the characters concerned, perhaps with a bespoke piece of creative dialogue. The key to balancing randomness and creative action is not to overdetermine the link from random stimuli to concrete outputs, but instead to view randomness as a high-level means of picking among different modes of low-level engagement.

To imagine what might happen were a writer to surrender even more control to a stochastic system such as the *I Ching* or D&D (or even Scéalextric) we need only look to Dick's novel within a novel. The "man in the high castle" of the title is a character named Hawthorne Abendsen, the enigmatic writer of a novel titled *The Grasshopper Lies Heavy* that offers an alternate history of World War II in which, shockingly, it was the Allies that won the war, turning Germany and Japan into client states. This alternate-alternate history serves as a beacon of hope for the people of the postwar United States, making Abendsen, its controversial writer, such a high-profile target that he is said to live in a fortress named the "high castle" in the Rocky Mountain zone. Abendsen's history is deeply at odds with the time line of Dick's novel, but it is also strangely different from the history of World War II as we all know it. The United States prevailed at Pearl Harbor because of the foresight of President Rex Tugwell, succeeding an FDR who, like Tiny Tim, did not die after all. It is Tugwell who ensures that the US fleet is not in port during the Japanese attack. Abendsen foresees the Allies falling to rancor among themselves after the war, with Britain winning a new cold war with the United States. However, he is not forthcoming when pressed on how he comes by his book's revelations, leaving it to his wife to admit his debt to the *I Ching*: "One by one Hawth made the choices. Thousands of them. By means of the lines. Historic period. Subject. Characters. Plot. It took years." She paints her husband as little more than the CPU that executed the plot-deciding algorithm of the *I Ching* to tell a tale that is at once both his and not his. Dick is being ironic, of course, as he gently mocks his own reliance on the *I Ching*, but his larger point is that history is just another story in which we are all mere "characters." Writers can tell such compelling stories using the mechanisms of simple chance because our own lives are subject to the very same mechanisms.

Users of the *I Ching* "consult the Oracle" with a specific question in mind. For instance, one of Dick's characters poses the question of whether Abendsen's novel is fiction or genuine history and generates the six lines of hexagram 61, *Chung Fu* ("Inner Truth"), with her coin tosses. With this, she infers that the novel is indeed factual and happily concludes that it is her own world that is fictional. But imagine a *Jeopardy!*-like version of bibliomancy, in which users generate the answer first and then find the question that fits this answer. In fact we *all* do this, insofar as reading makes bibliomancers of us all. Every time we read a book, or a news article, or a tweet, we cannot help but bring our own life experience to bear, to view the actions of another person or an imaginary figure as though they might be informed by, and inform in turn, our own issues and goals. Each new status update that pops into our Twitter time line invites us to see a relevance to our own lives. Although the *I Ching* would be a great topic for a Twitterbot—imagine a bot that delivers hexagrams and commentaries in response to any user who tweets the hashtag *#iChingMe*, using a sentiment analysis of recent tweets to guide its process of "oracular divination"—our bots already offer nuggets of thought-provoking text from a dynamic book that may be as specific as its own knowledge base or as general as the web or indeed all that language will allow. Consider the outputs of Parrish's *@everyword*, which tweeted every word of the English language in alphabetical sequence. Had users viewed its outputs as mere word listings, it would never have garnered the bulk of its seventy thousand followers. Even if many of its frequent lexical intrusions into our timeliness were ignored—and who can say that *every* word in the dictionary is worth tweeting?—it only takes a few percent of its outputs to attract our interest and stir our thoughts for such a bot to make a small but meaningful contribution to our day. These bots may not be able to read or distill "the tenor of a moment in the universe," but in their random exploration of the space of all possible signifiers for such moments, they create a potential for synchrony, in which a bot's outputs may occasionally (if quite accidentally) capture the mood of the Zeitgeist. So when *@everyword* tweeted "woman" on May 14, 2014, in the same week that the *New York Times* fired its first female executive editor for supposedly being "too bossy," the bot's followers may have felt that it was providing the real reason for Jill Abramson's dismissal. It requires a willing suspension of disbelief to think so, but it takes a comparable suspension of our most critical faculties to usefully engage with the *I Ching* too. This willing suspension is not willful ignorance, but a recognition of the value of nonliteral modes of expression and of nonliteral approaches to meaning.

The interviewer Charlie Rose offered this succinct analysis of the mainstream media's failure to predict a Trump victory in the 2016 US presidential election: "Those [on the left] who took him literally did not take him seriously, while those [on the right] who took him seriously did not take him literally." So the race was swayed less by those who believed Trump than by those who believed *in* him, with the businessman operating at, and thriving at, a level of nonliteralness that was unprecedented (or to use

out a victory on points, as the simplicity of templates gives them the edge over those whose messages rely on nuance, fact, and a willingness to see that complex problems often require complex solutions.

Twitter provides fertile ground for automated bluster and what has now come to be called *computational propaganda* by researchers who track the deleterious effects of bots on political discourse (the work of one such group of researchers can be found online at *politicalbots.org*). For instance, the use of hashtags can be as effective as a MAGA trucker hat in marking out the political leanings of a user, but just as hats cannot validate the true feelings of their wearers, at least outside the realm of Harry Potter and *@sortingbot*, hashtags are just as open to satirical use and abuse. Unfriendly agents may thus exploit and colonize the hashtags of their rivals, to insinuate themselves into their conversations under a false flag. In this way are proxy wars fought by our bots. Thus, supporters of Hillary Clinton in the 2016 presidential election cycle unleashed their bots to echo her battle cries and engage the supporters of her opponent, Donald Trump, on *their* territory, as marked out by their hashtags, and the supporters of Donald Trump did the same, bringing the battle to Hillary Clinton and her supporters via *their* hashtags. Hashtag "colonization"—say, when Clinton supporters use *#MAGA* for satirical effect or when Trump supporters use *#ImWithHer* to hurl brickbats at Clinton—devalues the hashtags of both sides, which is as good a reason as any for our bots not to use preexisting hashtags in their tweets unless they can bring something original and witty and openly bot generated to a conversation. Inevitably, when so many bots travel so widely on Twitter, bots of one political strain must often engage with those of another, like two zombies that shuffle around each other because each is unsure of the other's capacity to provide brainsss. We might find comfort in the idea that a real zombie apocalypse must eventually run its course as the supply of human brains dwindles, but zombies on Twitter may live forever (if Twitter lives on) by continuously feasting on the droppings of other zombies.

So are there protest and propaganda bots that are not zombies, and if so, how might we tell the difference between zombie and nonzombie in a way that is not self-serving? We need more than a codification of the view that "our bots are not zombies because they are ours; your bots are zombies because they are yours." We will find no hard criteria in which to anchor this ontological distinction, but we might as well begin with the idea of poetry. There is an urge toward creativity in the best of human protests, a reach for the figurative, the poetic, or the playful that shows the protester to be engaging at the level of ideas as well as utterances.

As good a taxonomy as any for politically charged bots can be found in Leonard Cohen's song "Bird on the Wire," for our bots are *our* birds on *our* wire, flying in circles with as much or as little altitude and speed as we care to give them.[18] Some, like *@arguetron*, are "like a worm on a hook" that baits a trap for hungry trolls, and others, like *@NRA_tally*, are "like a knight in an old-fashioned book," embodying a certain notion of social value. Yet bots like *@NRA_tally* and *@EveryTrumpDonor* preach mostly to the converted, with their very existence on Twitter mattering more than any particular tweet they might generate. It seems fair to infer that most followers of *@EveryTrumpDonor* are less interested in a $200 donation by a Texas dentist to Trump's presidential campaign than in the idea that any donor at all can be exposed to public scrutiny via the actions of a Twitterbot. These bots turn "following" into a political action, and the bots reward their followers with a sense of belonging and of having a tireless champion. Like the knight in Cohen's song, these bots proclaim, "I have saved all my ribbons for thee." Yet the most intriguing bots are also the least predictable, generating carefully packaged ideas that matter more than any single idea that the bot itself might embody. These bots strain against their guide wires, conveying at least the sense that they might occasionally transcend their limits and break free of our control. To repurpose Cohen's words, such a bot is "like a drunk in a midnight choir," but one that does more than screech another's distinctive words as an off-key caterwaul. Drunks sing their lines zestily, with an unhinged and irreverent inventiveness, especially if they forget, or never knew, how a line is supposed to go. In this irreverence and inventiveness lies the all-important power to surprise. All things considered, we should prefer our bots to act more like creative drunks than unthinking zombies.

Some bots are designed to tweet every day of the year, while others will be as seasonal as eggnog, green beer, or Cadbury crème eggs. So a Twitterbot offering Halloween costume suggestions might run from September to the end of October each year, while one that suggests offbeat gift ideas might tweet only in the run-up to Christmas. Let's suppose we set out to build a seasonal bot to poke fun at—or, more seriously, to protest—the official appointments of the new US president-elect, which in 2016 was Donald Trump. (If Hillary Clinton had won the election, our bot could just as easily take aim at her picks.) Such a bot will run for less than three months (November 9 to January 19) every four or eight years when a new president and a new administration takes power. Donald Trump's own Twitter-lofted kite flying regarding his appointments made this a timely topic for a bot at the end of 2016, with the president-elect's own tweets lending an unprecedented air of reality TV artifice to his putative picks. As speculation is just another form of invention that is partially grounded in fact, our bot can show a degree of creativity in individual tweets while parodying the selection process as a whole with its celebrity-obsessed modus operandi; that is, each specific tweet has an opportunity to make a reader think or laugh, or both, while the bot's lax grip on reality—as shown, for example, by its willingness to nominate fictional or dead people to important government positions—can serve to satirize Trump's real-life transition team's understanding of both their task and the nature of government.

As a minimal zombie baseline, we can start with a set of templates of the form "Trump taps [X] for Secretary of [S]," where [X] is a randomly chosen proper name from a list of candidate picks—we can use the NOC list as a source of famous names—and [S] names a government agency, again chosen randomly from a list that contains such staples as State, Treasury, Agriculture, Health and Commerce. Since any resonance between the fillers chosen for [X] and for [S] is going to be entirely accidental, this approach will do little to inspire a reader's confidence in the bot's understanding of its task, and if the bot works at all as a conceit, it will be because its random choice of fillers hints at the presumed randomness of the actual political process. "Look," it will effectively say to its followers, "the real process is as dumb as I am." But we can imbue the template-filling process with a bit more intelligence by exploiting the relational structure of the NOC to create a quirky mini-narrative with each tweet. So what we are aiming for here is the anarchic silliness of *Monty Python*'s famous "Ministry of Silly Walks" sketch, in which the bot invents a government department that is very likely absurd from the get-go, but then suggests an appointment to this absurdity that seems both silly yet somehow apt. For instance, the bot can create a separate department [S] for every Typical Activity in the NOC list, from "Running a Bureaucracy" to "Providing Comic Relief," and then fill [X] with the name of a famous person linked to that activity in the NOC, such as Adolf Eichmann for the former and Sideshow Bob or Baldrick for the latter. These pairings suggest that a president-elect is ignorant enough of government to believe that a department as silly as [S] exists, or should exist, yet is astute enough about cultural optics to pick the very best filler [X] that history, or fiction, has to offer. So our template now sheds a variable but takes on a more tightly knit internal structure as a result: "Trump taps [X] for Secretary of [X/Typical Activity]." Some instantiations of this template will, through no intention of the bot, strike more resonant notes than others, such as "Trump taps Ron Burgundy for

Secretary of Maintaining Salon-Quality Hair" and "Trump taps Vladimir Putin for Secretary of Bullying Neighboring Countries," yet even the silliest instantiations will present a scenario that is internally coherent and incongruously appropriate.

A diversity of templates can suggest different within-tweet scenarios that will be unpacked by readers to suggest varying cross-tweet narratives about a bot's real target. To satirize an incoming administration, it serves our goal for the bot to concoct scenarios that demonstrate a certain degree of informed fantasy on the part of the president-elect and his or her team. To use the phrase coined by Aristotle, it takes "educated insolence" to construct a fantasy informed by fact and conventional wisdom, even if a fantasy is crafted to showcase the presumed ignorance and stupidity of those whose worldview it is designed to satirize. Our templates can exploit knowledge of Group Affiliation in the NOC as follows (and, if space allows, the hashtag *#DrainTheSwamp* can be appended to the end of each):

Trump—who mocked [X/*Group Affiliation*] during the campaign—taps [X] for Secretary of [X/*Typical Activity*].

Trump, who once promised to shut down [X/*Group Affiliation*], picks [X] to lead Dept. of [X/*Typical Activity*].

Trump, who received millions from [X/*Group Affiliation*]'s PAC, picks [X] to lead Dept. of [X/*Typical Activity*].

Trump taps [X] for Dept. of [X/*Typical Activity*], despite FBI reports that [X/*Group Affiliation*] meddled in election.

Though [X/*Group Affiliation*] ran a Clinton SuperPAC, Trump taps [X] to head Dept. of [X/*Typical Activity*].

CIA says Russians have secret tape of Trump [X/*Typical Activity*] with [X]. Senate calls on [X/*Spouse*] to testify.

This might count as "fake news" if the scenarios painted in these tweets were not so inherently ridiculous, showing more kinship to the content of *The Onion* or *The Daily Show* than to that of *The Drudge Report* or *Breitbart News*. This may be storytelling tailored to a specific target and domain, but our bot's willingness to cross boundaries of fiction and history marks out its tales as informed nonsense. This template suggests an unhealthy mingling of reality with unreal "reality TV":

Wanting [X:*fictional*] for Sec. of [X/*Typical Activity*], Trump is told [X/*pronoun*] doesn't exist, picks [X/*Portrayed By*] instead.

This template turns a laudable intention into a risibly ineffective piece of theater:

Seeking to heal a divided nation, Trump nominates [X] and [X/Opponent] to jointly head Dept. of [X/Typical Activity].

The first can be instantiated as, "Wanting Jack Bauer for Sec. of Chasing Terrorists, Trump is told he doesn't exist, picks Keifer Sutherland instead," and the second as, "Seeking to heal a divided nation, Trump nominates Batman and The Joker to jointly head Dept. of Preventing Crime." We want our bot's stories to show insight into the foibles of their main character, a newly elected president, but they must also exaggerate the president's personality to signal their own counterfactuality. The following squeezes satire from a familiar narrative of one-upmanship:

Trump passes on [X] for Secretary of [X/Typical Activity], claiming to be the real brains behind [X/Creation].

This might be instantiated as "Trump passes on Al Gore for Secretary of Lecturing about Climate Change, claiming to be the real brains behind the Internet." Yet when a president's pick for head of the Environmental Protection Agency openly questions man-made climate change, there is a sense that no counterfactual could ever match the real thing for counterintuitive caprice. Reality takes on the hue of "you couldn't make this stuff up" when facts are paired so antagonistically as to create not harmony but friction. This is not zombie-like ignorance of the facts, but an impish disregard for facts so obvious the president *must* know them. Satire comes not from an ignorance of the facts but from a knowing disrespect for facts that are known to all or, in Aristotle's words, from educated insolence. Our insolent bots can satirically disrespect facts as willfully as any politician, and by dialing down the unhinged whimsy a little, we can magnify the satirical effect by channeling it via the lens of a few apropos facts. If education is knowledge and knowledge serves to constrain how a bot fills its templates, we can achieve a degree of educated insolence by building even more constraints into our bot's templates. These added constraints should steer the bot toward more intelligent picks that suggest an understanding of its task (e.g., secretaries of commerce should be experienced businesspeople, treasury secretaries are often rich) but they should not overconstrain it. They may constrain the choice of fillers to people in a specific taxonomic category or to those with specific Positive or Negative Talking Points, but because this is *meta*satire in action, we can still be surprised by how a bot chooses to fill our templates and satisfy *our* constraints:

Trump taps [*X* = *Businessman*] to be Secretary of Commerce, promises to make [*X/Typical Activities*] a priority.

Trump wants [*X* = *wealthy*] for Treasury Secretary, will make [*X/Typical Activities*] a first-term priority.

Trump promises to release his tax returns when [*X* = *wealthy*]—who made a bundle [*X/Typical Activities*]—does the same.

Trump taps [*X* = *media-savvy*] for White House Communications Chief, experience of [*X/Typical Activities*] considered valuable.

Trump appoints [*X* = *Criminal*] to head up the DOJ, brushes off a storied past of [*X/Typical Activities*].

Trump gifts Dept. of Energy to the energetic [*X* = *energetic*] as Sec. of [*X/Typical Activities*] already filled.

[*X* = *drug-addled*] to be Trump's pick for Surgeon General; dealer connections considered a plus.

Fighting fire with fire, Trump appoints [*X* = *ruthless*] as counterterrorism advisor, experience [*X/Typical Activities*] a plus.

To promote the American dream, Trump appoints [*X* = *inspiring*] to be Secretary of State, looks forward to [*X/Typical Activities*] together.

If these constraints provide the "educated" side of the bargain, their obvious inadequacy as a filter for whimsy and absurdity provides the "insolence." Though it is perfectly reasonable for a president to pick a business leader for the position of secretary of commerce, and to make that person's business goals his own, our concrete choices often mock our generic aspirations, as in these instances: "Trump taps Ebenezer Scrooge to be Secretary of Commerce, promises to make pinching pennies a priority" and, "Trump wants Lex Luthor for Treasury Secretary, will make promoting greed a first-term priority." Imagine if every mindless use of formulaic language could be exploded from within like this! Well, when a disloyal friend whines, "We were like brothers once," you can always retort, "Yes, Cain and Abel," or, "I know, Michael and Fredo Corleone." If an employee seeks a raise with the dubious claim that "I do the work of two people for this company," you can always reply, with analogical righteousness, "Yes, Laurel and Hardy." Or if an angry spouse points to a mob of dung-flinging apes on the TV and mutters, "Your relatives, no doubt," you might hope for the words, "Yes, my in-laws," to trip off your tongue. Formulaic templates assume equally formulaic fillers, such as loving brothers, hard workers, and blood relatives, not the category outliers of brothers who kill

to the same Twitter drumbeat. So truth rubs elbows with half-truths, lies, and Frankfurt-style "bullshit," while dreams and fantasies mingle with hard reality.

Bots such as @pentametron contrive to reorder the pages of the Twitter book so that mundane wakefulness becomes poetic dreaming, by nudging readers to perceive a semantic or pragmatic resonance between tweets that are paired on metrical grounds only. It works much like a dating agency that pairs members on the principle that it is the couples with similar height, weight, or other superficial measures of compatibility that generate the most electrifying sparks. In this way, the bot adds value to tweets that may have already run their authors' intended courses. Indeed, by striking sparks from what might otherwise seem like spent fuel, our bots confront head-on the specter of disposability that haunts not just Twitter but older forms of content delivery too. The cultural critic Mark Lawson, who writes extensively about television, diagnosed TV's pre-HBO lack of artistic standing with an insight that seems as relevant to Twitter now as to TV in the 1980s and 1990s: "The invisibility to posterity has always been television's difficulty. Many programs are intended to be disposable, to disintegrate even as you look at them."[21] Even the wittiest tweets are disposable ephemera, flashes of light that quickly recede as our time lines fill with new content. Our bots only quicken the pace with which they recede from view and from memory by using automation to ensure that new content is produced with clockwork regularity to supplant the old. Our bots can no more hold back time than King Canute could hold back the tides, but that has never been their purpose. Bots can give new life even to content that is intended to be disposable and take from their disintegration the material for new tweets. Moreover, as our Twitter lathes produce their steady streams of sparks in the form of whimsical what-ifs, these may in turn ignite the imaginations of human users (and perhaps other bots) who might then refine, repackage, or satirize these ideas in a never-ending cycle of disposable creativity. Individual tweets may be disposable, but the overall cycle of creativity lives on.

## Twitter Toys Last All Summer Long

We build our Twitterbots to be tourists in strange lands. We set them loose to explore those pocket universes on our behalf and to send us frequent postcards on what they see there. These are realms of pure imagination, not of hard reality, but they are worlds that often mock our own with their simplicity, freedom, and elevation of form over meaning. Fans of TV's *Rick and Morty* might see parallels here with Rick's interdimensional cable box that allows him and the Smith family to watch an infinitude of inventively awful (yet bizarrely attention-holding) shows from across the multiverse. Here is the show's introduction to an oddly familiar *Saturday Night Live!* that is wildly popular on another world:

It's *Saturday Night Live!* Starring a piece of toast, two guys with handlebar mustaches, a man painted silver who makes robot noises, Garmanarar, three s-eh-bl-um-uh-uh-uh- I'll get back to that one, a hole in the wall where the men can see it all, and returning for his twenty-fifth consecutive year, Bobby Moynihan![22]

This may sound like TV made by Twitterbots, but who wouldn't want to channel-surf shows like these? Other briefly glimpsed shows from skew-whiff universes of S1E8 include a *Games of Thrones*, where everyone is a dwarf, except, of course, for the vertically challenged Tyrion Lannister, and a poorly paced true crime show, *Quick Mysteries*, that reveals all of its cards up front. The rapid-fire invention of *Rick and Morty* reminds us that we humans are the universal what-if machine, the mental equivalent of Rick's interdimensional cable box, capable of inventing endless new worlds to visit. Though our Twitterbots may be far from universal, we can build them to explore bespoke new worlds on our behalf, to dig deeper than our schematic view of the world and its rules might otherwise allow, to ferret out the weirdest instantiations of these rules for us to sample and enjoy. So while our individual bots may resemble a Bizarro channel on multiverse TV, with its weird tics, whacky obsessions, and view-askew take on life—for instance, despite the seriousness of @NRA_Tally, its four-hourly killing sprees can read like a deliberate parody of modern cable programming—collectively they turn Twitter into Rick's interdimensional cable box, allowing us all to channel-surf the wonders of a multiverse where humans are just one voice among many.

The various resources described in this book have been designed to allow bot builders to respond nimbly to new what-if opportunities for Twitterbots as the holidays, the seasons, and changing circumstances present them. They allow our Twitterbots to bring a quirky understanding of this world to their automated explorations of other worlds, to help them appreciate what they find there, and to help them to better filter the noisome chaff from the tweet-worthy wheat. By using knowledge, scant though it may be, to lend some familiarity to the oddities of an artificial world, they also inevitably show us the strangeness of our own. This is what Twitterbots do best: they remind us of the strangeness of language, social convention, and human nature more generally by allowing us to see familiar human qualities in the synthetic, the mechanical, and the alien. In your onward explorations of this multiverse of possibilities, do consider docking occasionally at *BestOfBotWorlds.com*, to share your experiences with others and to refuel on resources and ideas. In the final analysis, it is not the Twitterbots but the Twitterbot builders that make Twitter the best of bot worlds.

## Trace Elements

In the spirit of opening new doors while closing others, we conclude this final chapter by building the ultimate what-if machine: an interdimensional cable box of our own. We can glimpse the possibility space of television in the tweets of a restless channel-hopping Twitterbot, which we will build by repurposing a variety of generative components from the Tracery grammars of other bots in earlier chapters. As good a place as any to start the construction of our bot is a mainstay of the TV viewing experience in all dimensions: advertising.

One of our earliest Tracery grammars from chapter 3 exploited the power of raw combinatorial generation to coin new words and new meanings from the collision of Greek and Latin roots and their standard interpretations. As the classical roots of these new words are often suggestive of the kinds of products we might discreetly seek out in a pharmacy, our neologism grammar is easily converted into a generator of faux-scientific gizmos and doodads. Our product pitches will make for more compelling TV if we recruit the famous faces of the NOC list to act as celebrity shills, as in this commercial reframing:

Ernst Stavro Blofeld swears by "GaleoMart." When you need a place of business dedicated to the sale of sharks, there's none better!

This pairing of sharks with a Bond villain is merely an accident of random generation, yet we build our bots to foster such happy accidents. To cultivate many more combinatorial delights, we can repurpose our *Just Desserts* grammar from chapter 3. While those treats were created to be vengefully vile, perhaps this is how the denizens of other worlds actually prefer their desserts. By also defining a set of apt prefixes and suffixes to combine, we can generate names for the companies that make the awful treats, as in this grammar output:

Why not try TrumpDessertz-brand Peanut butter cups made with used coffee grounds instead of chocolate—Feed your desires!

Famous people can also provide celebrity endorsements for the services of a company with deep pockets, and a diverse source of possible services is to be found in the moral maze of action frames and roles we explored in chapter 4. We can repurpose our moral grammar to generate advertising such as this:

When Lord Voldemort wants to commit top-notch killing he calls 555-Predators. They won't be beaten on price

In chapter 7 we built a grammar to map the normative properties of familiar visual ideas to their related dimensions. This allowed us, for example, to pitch the dull gray of solid rock as the representative color of solidity itself. The same approach can be used to lend a familiar face to more abstract dimensions, as when we reinvent a NOC talking point as a celebrity perfume:

Parfum de "Reclusiveness"—the new scent for men from JD Salinger.

These four generators combine to form a single Tracery grammar named *Advertising grammar.txt* in the Interdimensional Cable directory of the TraceElements repository. This grammar provides the foundations for the rest of our dimension-hopping TV service, on which we will now build another fixture of the cable landscape: the twenty-four-hour news cycle. Any news service needs newsworthy propositions to broadcast, and even assertions about the most famous people need interesting claims at their core to reach the air. Darius Kazemi's @twoheadlines is ideal in this respect, as it is constantly fueled by up-to-the-minute headlines on the web. However, to squeeze our news generator into a Tracery bot, we shall have to confine ourselves to a closed-world model of newsworthiness. Fortunately, as we saw in chapter 9, dbpedia is a rich source of conversation-worthy categorizations that can be as informative as good gossip and headlines. By harvesting every dbpedia category that matches any of the patterns X_who_have_Y, X_who_were_Y, X_who_can_Y and X_about_Y we can extract the central claims (the Ys) around which our bot's news headlines will be based. It then remains for our bot to invent a suitable name for the cable news network in question. Using Fox News as our inspiring exemplar, we leverage our list of animals from chapter 3 to create other animalistic news stations for the multiverse, as in:

JaguarNews Exclusive: Vladimir Putin has nothing nice to say about The Irish Mob. Up Next ...

MonkeyNews Exclusive: Martha Stewart denies claims she has acquired Austrian citizenship. Stay tuned

25. Writing in "The Intersect" for the *Washington Post* in 2014, journalist and bot builder Caitlin Dewey noted that "the idea of using Twitter as a medium for serious art and social commentary has increasingly caught on with a ragtag group of conceptual writers, generative poets, and performance artists": https://washingtonpost .com/news/the-intersect/wp/2014/05/23/what-happens-when-everyword-ends.

26. The screenplay for the 1999 movie *Mystery Men* was written by Neil Cuthbert and Bob Burden.

27. The discussion of cocktail party (or chatterbox) syndrome cites this paper in particular: Ellen R. Schwartz, "Characteristics of Speech and Language Development in the Child with Myelomeningocele and Hydrocephalus," *Journal of Speech and Hearing Disorders* 39 (1974): 465.

28. The paper cited on cocktail party syndrome is Neil P. McKeganey, "The Cocktail-Party Syndrome," *Journal of Sociology of Health and Illness* 5, no. 1 (1983): 95–103.

29. Alan M. Turing, "Computing Machinery and Intelligence," *Mind* 59 (1950): 433–460.

30. This quote from George Orwell comes from his 1945 essay, "Funny, But Not Vulgar," reprinted in George Orwell, *The Collected Essays, Journalism and Letters of George Orwell* (New York: Harcourt, 1968). The full text of the essay is online as well: http://orwell.ru/library/articles/funny/english/e_funny.

31. Ada Lovelace and her ideas on a new poetical science are discussed in Betty A. Toole, *Ada, the Enchantress of Numbers: Prophet of the Computer Age* (Moreton-in-Marsh, UK: Strawberry Press, 1998).

## Chapter 2: The Best of Bot Worlds

1. The Collection of the Metropolitan Museum of Art can be accessed online at http://www.metmuseum.org/art/collection.

2. The script for the sketch and an online video can be accessed online at http://abitoffryandlaurie.co.uk/sketches/language_conversation.

3. Noam Chomsky's most memorable nonce sentence was "colorless green ideas sleep furiously." Noam Chomsky, *Syntactic Structures* (The Hague; Paris: Mouton, 1957).

4. The philosophical concept of the sublime is discussed at length in Emily Brady, *The Sublime in Modern Philosophy: Aesthetics, Ethics, and Nature* (Cambridge: Cambridge University Press, 2013).

5. *wikiHow: How to Do Anything*. The website is http://www.wikihow.com.

6. Joseph Weizenbaum's 1966 ELIZA paper is a true classic and a real joy to read. Joseph Weizenbaum, "ELIZA—A Computer Program for the Study of Natural

Language Communication between Man and Machine," *Communications of the ACM* 9, no. 1 (1966): 36–45. A scan can be downloaded here: http://dl.acm.org/citation .cfm?id=365168.

7. There are several interactive versions of ELIZA that readers can play with online. These are accessible links for it on the corresponding Wikipedia page: https:// en.wikipedia.org/wiki/ELIZA.

8. Weizenbaum wrote about ELIZA in cautionary terms that reflect a growing concern about AI: Joseph Weizenbaum, *Computer Power and Human Reason: From Judgment to Calculation* (New York: Freeman, 1976).

9. Daniel Shiffman offers a highly informative introduction to Markov text genera-tion (MTG) on his website: http://shiffman.net/a2z/markov/.

10. http://www.bearstearnsbravo.com/.

11. This excerpt from a speech by William S. Burroughs can be heard on the 1999 album *Break Out in Grey Room* (on the Sub Rosa label), which brings together records of various Burroughs speeches.

12. The screenplay for the 2005 movie *Stealth* was written by W. D. Richter.

13. The British newspaper *Daily Mail* reported on some of Tay's troubling tweets before they were deleted. They allegedly included such horrors as "Bush did 9/11 and Hitler would have done a better job than the monkey we have got now." http:// www.dailymail.co.uk/sciencetech/article-3507826/Tay-teenage-AI-goes-rails-Twitter -bot-starts-posting-offensive-racist-comments-just-day-launching.html.

14. https://www.reddit.com/user/astro-bot/. For the "coolest" label. see https:// www.reddit.com/r/AskReddit/comments/4hwm38/what_is_your_favourite_reddit _bot/d2t3h1t.

15. Darius Kazemi, "Basic Twitter Bot Etiquette," *Tiny Subversions*, March 16, 2013, http://tinysubversions.com/2013/03/basic-twitter-bot-etiquette/.

16. "Bots Should Punch Up," *News You Can Bruise*, November 27, 2013, https:// www.crummy.com/2013/11/27/0.

**Chapter 3: Make Something That Makes Something**

1. "Pub names," Wikipedia, last modified November 2, 2017, 06:45, https://en. wikipedia.org/wiki/Pub_names.

2. A key reference to this AI technology, which marries syntax and semantics in a single grammar formalism, is Richard R. Burton, "Semantic Grammar: An Engineer-ing Technique for Constructing Natural Language Understanding Systems," *ACM SIGART Bulletin* no. 61 (1977): 26.

11. Rob Dubbin discussed his *@RealHumanPraise* in "The Rise of Twitterbots," *New Yorker*, November 14, 2013, http://www.newyorker.com/tech/elements/the-rise-of-twitter-bots.

12. An emoji is a pictographic character of a kind that is now commonplace in digital communications, texting, tweeting, and more. The website http://emoji-tracker.com/ (courtesy of Matthew Rothenberg) tracks the use of different emoji in real time and presents the results in a rapidly changing scoreboard. Rothenberg's article on the genesis of EmojiTracker provides a wealth of detail on emoji themselves: https://medium.com/@mroth/how-i-built-emojitracker-179cfd8238ac.

13. Matthew Rothenberg's *@EmojiDoll* Twitterbot generates combinations of emoji characters in a doll-like configuration to represent any Twitter user that requests the EmojiDoll experience. An article by Lucia Peters for *Bustle* in July 2014 explores the bot and its oeuvre: https://www.bustle.com/articles/31579-emojidoll-twitter-bot-draws-surprisingly-accurate-portraits-of-our-innermost-selves-completely-in-emojis.

14. The various houses of the fictional school Hogwarts are described in detail on a wiki devoted to J. K. Rowling's creations: http://harrypotter.wikia.com/wiki/Hogwarts_Houses.

15. The author of the *Harry Potter* novels, J. K. Rowling, describes the genesis of the sorting hat in a blog post for *Pottermore*: https://www.pottermore.com/writing-by-jk-rowling/the-sorting-hat. She notes that her first idea for a sorting mechanism was not a magical hat but an overly complex "Heath Robinson–ish" (or Rube Goldberg) machine.

16. Darius Kazemi describes the genesis of his bot *@SortingBot* on his website: http://tinysubversions.com/notes/sorting-bot/.

17. http://www.newyorker.com/cartoons/bob-mankoff/graphs-and-laughs.

18. The 2016 Cognitive Science Society conference was held at the Philadelphia Convention Center in August 2016; its theme was "Integrating Psychological, Philosophical, Linguistic, Computational and Neural Perspectives."

### Chapter 7: Magic Carpets

1. The screenplay for the 1998 film *The Big Lebowski* by the Coen brothers can be accessed here: http://www.inwardeleven.com/lebowski/.

2. "The Circus Animals' Desertion" was among the last poems written by W. B. Yeats and was published, fittingly, in a collection entitled *Last Poems* in 1939. The full text of the poem can be read here: https://www.poetryfoundation.org/poems-and-poets/poems/detail/43299.

3. Andi McClure and Michael Brough, the creators of the game *Become a Great Artist in 10 Seconds*, spoke to *GamaSutra* in 2015 about their creation: http://www.gama sutra.com/view/news/237027/Road_to_the_IGF_McClure_and_Broughs_Become_a _Great_Artist_in_Just_10_Seconds.php.

4. Allison Parrish talks about the genesis of the *@the_ephemerides* Twitterbot (and about related projects and ideas too) on http://www.decontextualize.com/.

5. Susie Hodge, *Why Your Five-Year-Old Could Not Have Done That* (London: Thames and Hudson, 2012).

6. Breton encountered this enigmatic phrase in Lautréamont's book *Les Chants de Maldoror*, first published in 1874. The book shares many elements with modern Twitterbots, and the phrase in question came from a section in which the author completes the simile "as beautiful as …" in a variety of surprising but resonant ways.

7. The lyrics to Burl Ives's "The Lollipop Tree" can be found here (with a link to a YouTube video): http://www.streetdirectory.com/lyricadvisor/song/ulwuu/the_lolli pop_tree/.

8. Thorsten Brants and Alex Franz, *Web IT 5-Gram Database, Version 1* (Philadelphia: Linguistic Data Consortium, 2006).

9. The name *The Game of Life* was coined by mathematician John H. Conway for a cellular automaton. The automaton, and the name, were later popularized in Martin Gardner, "The Fantastic Combinations of John Conway's New Solitaire Game 'Life,'" *Scientific American* (October 1970). A copy of the article can be accessed via this link: http://ddi.cs.uni-potsdam.de/HyFISCH/Produzieren/lis_projekt/proj_gamelife/ ConwayScientificAmerican.htm.

10. A listing of the many varieties of emergent structure in Conway's *Game of Life* can be found at http://www.conwaylife.com/wiki.

11. The compact numbering system for the rules of elementary cellular automata, and for the automata themselves, is described in Stephen Wolfram, "Statistical Mechanics of Cellular Automata," *Reviews of Modern Physics* 55 (1983): 601–644, and his *A New Kind of Science* (Champaign, IL: Wolfram Media, 2002).

12. Lévy processes are named for the French mathematician Paul Pierre Lévy, though the specific notion of a "Lévy flight" was coined by another mathematician, Benoit Mandelbrot, in his seminal book: Benoit B. Mandelbrot, *The Fractal Geometry of Nature* (New York: Freeman, 1982).

13. David Perkin elaborates the conceptual metaphor of an abstract search space into a gold-rich Klondike space in *The Eureka Effect: The Art and Logic of Breakthrough Thinking* (New York: Norton, 2001).

14. The Twitter status (and a photo-realistic image of the corresponding moth from *@mothgenerator*) can be found at https://twitter.com/mothgenerator/status/779235998207213568.

15. A good introduction to semiotics is offered by Daniel Chandler, *Semiotics: The Basics* (London: Routledge, 2007) Alternately, Chandler provides an old-school HTML course on semiotics at http://visual-memory.co.uk/daniel/Documents/S4B/.

16. Ben Zimmer recounted the history of Colbert's comedic coinage "Truthiness" in his article "Truthiness," *New York Times Magazine,* October 13, 2010, http://www.nytimes.com/2010/10/17/magazine/17FOB-onlanguage-t.html.

17. Hjelmslev's framework is discussed in Chandler's *Semiotics.*

18. "A tweet is a linguistic container much like any other." This idea is elaborated at length in Tony Veale, "The Shape of Tweets to Come: Automating Language Play in Social Networks," in Nancy Bell, ed., *Multiple Perspectives on Language Play* (Boston: Mouton DeGruyter, 2016), 73–92.

### Chapter 8: Bot-Time Stories

1. The story of how Jean-Luc Godard's 1965 genre-bending film *Alphaville* was almost called *Tarzan versus IBM* is recounted in Chris Darke, *Alphaville (Jean-Luc Godard, 1965)* (London: I. B. Tauris, 2005).

2. The term *hypertext* first appeared in this technical paper: Theodor H. Nelson, "Complex Information Processing: A File Structure for the Complex, the Changing, and the Indeterminate," in *Proceedings of the National Conference of the Association for Computing Machinery* (New York: ACM, 1965).

3. Flann O'Brien, *At Swim-Two-Birds* (Dublin, Ireland: Dalkey Archive, 1939).

4. The twelve-part comic book series *Watchmen* was written by Alan Moore and drawn by Dave Gibbons and first published by DC Comics in 1986 and 1987. An oral history of *Watchmen* has been created by *Entertainment Weekly*: http://ew.com/article/2005/10/21/watchmen-oral-history/. The comic book series *The League of Extraordinary Gentlemen* was written by Alan Moore and drawn by Kevin O'Neill. It was first published by an imprint of DC Comics in 1999.

5. Raymond Chandler, *The Big Sleep* (New York: Knopf, 1939).

6. Philip K. Dick, *Do Androids Dream of Electric Sheep?* (New York: Doubleday, 1968). It formed the basis for the 1982 movie *Blade Runner*, directed by Ridley Scott from a script by Hampton Fancher and David Peoples.

7. The 1962 film *Creation of the Humanoids* predates *Blade Runner* by twenty years and Philip K. Dick's source novel by six years. It is a film with a highly quotable script that supports many repeat viewings and drinking games.

8. The musical *Into the Woods*, with music and lyrics by Stephen Sondheim, premiered on Broadway in 1987. It was turned into a movie by Walt Disney studios in 2014.

9. John Yorke, *Into the Woods: A Five Act Journey into Story* (London: Penguin Books, 2014). The book sees all stories as journeys of discovery and change, and sees itself as a journey, too.

10. Joseph Campbell's book revealed the deep structure of heroic myth, and of heroic stories more generally. Joseph Campbell, *The Hero with a Thousand Faces* (New York: Pantheon Books, 1949).

11. Christopher Vogler's original memo, with commentary by the author, is available online: http://www.thewritersjourney.com/hero's_journey.htm#Memo.

12. The script of the 1999 movie *The Matrix* by Lana and Lilly Wachowski can be viewed at http://www.imsdb.com/scripts/Matrix,-The.html.

13. The 1996 film *Scream* was written by Kevin Williamson and directed by Wes Craven. The slasher movie "rules" that are ironically mocked in Scream are enumerated on its wiki: http://scream.wikia.com/wiki/The_Rules.

14. Hitchcock coined the nonce term *MacGuffin* in a speech at Columbia University and later used it in an interview with François Truffaut for this book: Francois Truffaut, *Hitchcock* (New York: Simon and Schuster, 1967).

15. Vladimir Propp's book was written in Russian and later translated into English for an international audience: Vladimir Propp, *Morphology of the Folktale* (Bloomington, IN: American Folklore Society, 1957).

16. Stith Thompson has written extensively on the structure of folktales—see, for example, Stith Thompson, *The Folktale* (Berkeley: University of California Press, 1977).

17. *Law and Order* first aired in 1990. Each week it took viewers on a serpentine journey from murder scene to courtroom with plots that had their own distinctive Propp-like structure.

18. William Wallace Cook's book was first published in 1928 by the Ellis Publishing Company and has since been reprinted by Tin House Books. William Wallace Cook, *Plotto: The Master Book of All Plots* (Portland, OR: Tin House Books, 2011).

19. William S. Burroughs and his collaborator Brion Gysin popularized the cut-up method in the 1960s. William S. Burroughs, "The Cut-Up Method," in Leroi Jones, ed., *The Moderns: An Anthology of New Writing in America* (New York: Corinth Books, 1963).

20. *Wikipedia.org* is a well-known and respected source of encyclopedic knowledge in textual form. Developers seeking a more structured source of knowledge can turn

# Subject Index

# Bot Index